Integrated Pest Management in Horticulture

Integrated Pest Management in Horticulture

Guest Editors

Małgorzata Tartanus
Eligio Malusà

Basel • Beijing • Wuhan • Barcelona • Belgrade • Novi Sad • Cluj • Manchester

Guest Editors

Małgorzata Tartanus
Department of Plant
Protection
The National Institute of
Horticultural Research
Skierniewice
Poland

Eligio Malusà
Department of Plant
Protection
The National Institute of
Horticultural Research
Skierniewice
Poland

Editorial Office
MDPI AG
Grosspeteranlage 5
4052 Basel, Switzerland

This is a reprint of the Special Issue, published open access by the journal *Horticulturae* (ISSN 2311-7524), freely accessible at: https://www.mdpi.com/journal/horticulturae/special_issues/IPM_Horticulture.

For citation purposes, cite each article independently as indicated on the article page online and as indicated below:

Lastname, A.A.; Lastname, B.B. Article Title. *Journal Name* **Year**, *Volume Number*, Page Range.

ISBN 978-3-7258-3709-0 (Hbk)
ISBN 978-3-7258-3710-6 (PDF)
https://doi.org/10.3390/books978-3-7258-3710-6

Cover image courtesy of Małgorzata Tartanus

© 2025 by the authors. Articles in this book are Open Access and distributed under the Creative Commons Attribution (CC BY) license. The book as a whole is distributed by MDPI under the terms and conditions of the Creative Commons Attribution-NonCommercial-NoDerivs (CC BY-NC-ND) license (https://creativecommons.org/licenses/by-nc-nd/4.0/).

Contents

Małgorzata Tartanus and Eligio Malusà
Drivers of and Barriers to the Implementation of Integrated Pest Management in Horticultural Crops
Reprinted from: *Horticulturae* 2024, 10, 626, https://doi.org/10.3390/horticulturae10060626 . . . 1

Syed Arif Hussain Rizvi, Laila A. Al-Shuraym, Mariam S. Al-Ghamdi, Fahd Mohammed Abd Al Galil, Fahd A. Al-Mekhlafi, Mohamed Wadaan and Waqar Jaleel
Evaluation of *Artemisia absinthium* L. Essential Oil as a Potential Novel Prophylactic against the Asian Citrus Psyllid *Diaphorina citri* Kuwayama
Reprinted from: *Horticulturae* 2023, 9, 758, https://doi.org/10.3390/horticulturae9070758 4

José Alfonso Gómez-Guzmán, José M. Herrera, Vanesa Rivera, Sílvia Barreiro, José Muñoz-Rojas, Roberto García-Ruiz and Ramón González-Ruiz
A Comparison of IPM and Organic Farming Systems Based on the Efficiency of Oophagous Predation on the Olive Moth (*Prays oleae* Bernard) in Olive Groves of Southern Iberia
Reprinted from: *Horticulturae* 2022, 8, 977, https://doi.org/10.3390/horticulturae8100977 18

Randa Mahmoudi, Malik Laamari and Arturo Goldarazena
Assessment of Thrips Diversity Associated with Two Olive Varieties (Chemlal & Sigoise), in Northeast Algeria
Reprinted from: *Horticulturae* 2023, 9, 107, https://doi.org/10.3390/horticulturae9010107 32

András Lajos Juhász, Márk Szalai and Ágnes Szénási
Assessing the Impact of Variety, Irrigation, and Plant Distance on Predatory and Phytophagous Insects in Chili
Reprinted from: *Horticulturae* 2022, 8, 741, https://doi.org/10.3390/horticulturae8080741 43

Fani Fauziah, Agus Dana Permana and Ahmad Faizal
Characterization of Volatile Compounds from Tea Plants (*Camellia sinensis* (L.) Kuntze) and the Effect of Identified Compounds on *Empoasca flavescens* Behavior
Reprinted from: *Horticulturae* 2022, 8, 623, https://doi.org/10.3390/horticulturae8070623 55

Ahmed Soliman, Saleh Matar and Gaber Abo-Zaid
Production of *Bacillus velezensis* Strain GB1 as a Biocontrol Agent and Its Impact on *Bemisia tabaci* by Inducing Systemic Resistance in a Squash Plant
Reprinted from: *Horticulturae* 2022, 8, 511, https://doi.org/10.3390/horticulturae8060511 67

Secilia E. Mrosso, Patrick Alois Ndakidemi and Ernest R. Mbega
Farmers' Knowledge on Whitefly Populousness among Tomato Insect Pests and Their Management Options in Tomato in Tanzania
Reprinted from: *Horticulturae* 2023, 9, 253, https://doi.org/10.3390/horticulturae9020253 83

Małgorzata Tartanus, Barbara Sobieszek, Agnieszka Furmańczyk-Gnyp and Eligio Malusà
Integrated Control of Scales on Highbush Blueberry in Poland
Reprinted from: *Horticulturae* 2023, 9, 604, https://doi.org/10.3390/horticulturae9050604 97

Lauren Helen Farwell, Greg Deakin, Adrian Lee Harris, Georgina Fagg, Thomas Passey, Carol Verheecke-Vaessen, et al.
Cladosporium Species: The Predominant Species Present on Raspberries from the U.K. and Spain and Their Ability to Cause Skin and Stigmata Infections
Reprinted from: *Horticulturae* 2023, 9, 128, https://doi.org/10.3390/horticulturae9020128 111

Héctor G. Núñez-Palenius, Blanca E. Orosco-Alcalá, Isidro Espitia-Vázquez, Víctor Olalde-Portugal, Mariana Hoflack-Culebro, Luis F. Ramírez-Santoyo, et al.
Biological Control of Downy Mildew and Yield Enhancement of Cucumber Plants by *Trichoderma harzianum* and *Bacillus subtilis* (Ehrenberg) under Greenhouse Conditions
Reprinted from: *Horticulturae* **2022**, *8*, 1133, https://doi.org/10.3390/horticulturae8121133 . . . **122**

Editorial

Drivers of and Barriers to the Implementation of Integrated Pest Management in Horticultural Crops

Małgorzata Tartanus and Eligio Malusà *

The National Institute of Horticultural Research, 96-100 Skierniewice, Poland; malgorzata.tartanus@inhort.pl
* Correspondence: eligio.malusa@inhort.pl

Integrated pest management (IPM) aims to protect plants using methods that limit the use of pesticides, as well as other interventions, to levels that are economically and ecologically justified, thus reducing the negative impact of crop protection on humans and the environment [1]. The adoption of IPM has been successful in fruit and protected crops, but remains marginal in other crops [2].

The routine monitoring of pests, the application of pest population thresholds, and the rational use of pesticides are the major pillars upon which IPM is based. Adequate pest monitoring systems using innovative tools, such as automated insect/spore traps [3] or forecasting systems [4], are currently available and are critical for the correct implementation of IPM. Tools predicting the dynamics of harmful organisms, including epidemiological models, have been developed to provide decision support systems (DSSs) that integrate intervention thresholds, and should thus promote the implementation of IPM programs that are adaptable to specific needs [5,6].

Biological control (BC) is a key ecosystem service and can become a pillar of IPM, for example, in case of protected crops as well as the control of invasive pests through the introduction of their natural enemies. Indeed, climate change as well as intensification of trade have favored the introduction of new invasive pests and their establishment in new territories on a global scale [7]. *Drosophila suzukii* [8] and *Hyalomorpha hyalis* [9] are two recent examples of this trend and the potential of BC. Nevertheless, while it was estimated that approximately 230 natural enemies are available from over 500 suppliers [10], biological control is still adopted on a relatively small scale world-wide. Reasons for this include the need for technical knowledge about the conditions for the successful large-scale release of natural enemies (e.g., the determination of the opportune time for their release, when there is an abundance of the target pest population on which the predator or parasitoid feeds) as well as the ecological and behavioral aspects of the natural enemy [11,12].

On the other hand, the exploitation of conservation- BC and crop management practices to increase the abundance and activity of autochthonous natural enemies is becoming increasingly relevant considering the decrease in biodiversity recorded in recent decades [13]. In this regard, landscape modification at the farm [14] and/or territorial level [15] has had a favorable impact on the whole control strategy and positive long-term results have also demonstrated favorable economic returns on investment [16]. Shelters or food sources for natural enemies can be provided by specific plant structures, groups of plants, or plant litter, as well as through artificial structures. Interconnectivity among these structures could improve the movement of natural enemies within and between fields [17], with large distances between these structures affecting the efficiency of BC [18].

Understanding the economic value of the key ecosystem services supplied by BC could encourage the adoption of IPM practices and increase awareness among decision makers supporting funding measures. The relationship between the composition of the surrounding landscape and biodiversity in crop fields has been found to impact ecosystem services (e.g., pollination) as well as pest regulation through conservative BC [19,20]. For example, an estimation of the monetary value of ecosystem services derived from the

Citation: Tartanus, M.; Malusà, E. Drivers of and Barriers to the Implementation of Integrated Pest Management in Horticultural Crops. *Horticulturae* **2024**, *10*, 626. https://doi.org/10.3390/horticulturae10060626

Received: 14 May 2024
Revised: 23 May 2024
Accepted: 31 May 2024
Published: 12 June 2024

Copyright: © 2024 by the authors. Licensee MDPI, Basel, Switzerland. This article is an open access article distributed under the terms and conditions of the Creative Commons Attribution (CC BY) license (https://creativecommons.org/licenses/by/4.0/).

introduction of living mulches, a practice commonly recommended to increase biodiversity in orchards, demonstrated the potential for additional income also due to the reduced need for pest control treatments [21]. The latter is particularly relevant considering that pest management decisions can generate externalities (i.e., costs or benefits that are not directly reflected in economic transactions), such as long-term effects on environmental health, water quality, or workers' health, with ecosystem-level estimates showing significant value [22]. However, if growers cannot directly benefit from the externalities provided by BC, they may under-adopt IPM practices. Considering that the economic threshold is a key component of IPM, which incorporates population dynamics into an economically based framework supporting control decisions, the limited research evaluating BC or conservative BC makes difficult to convince farmers to adopt such practices [23]. In this case, public support could help promote IPM implementation. Nevertheless, despite the strong policy support for IPM, particularly in advanced economies in the last few decades, its adoption has been inconsistent [24,25].

Although most technological constraints representing potential obstacles to the adoption of IPM have been overcome by recent technological progress (i.e., image analysis for insect identification, meteorological data acquisition, internet access, simple phone-based applications, etc.) [26], the translation of these innovative methods into practice is still hampered by socio-economic constraints and farmers' perceptions of IPM being a complex, technically demanding practice, not always ensuring economic advantages with respect to conventional or organic farming [27,28].

Nevertheless, the introduction of precision farming techniques (which can reduce the input of PPPs only to areas of fields where the economic threshold is reached) in combination with artificial intelligence could also represent an important technological innovation in terms of reduced costs [29]. Moreover, the reduction in environmental impact achieved through the application of pesticides using spraying systems, which allows farmers to recover product that has not reached the plant canopy or dynamically modifying the amount applied to different parts of the tree canopy using sensors and IT systems in their tractors, are innovations also associated with economic benefit [30]. Nonetheless, the efficiency of precision farming systems depends on the accuracy of the monitoring method selected for the specific pest, as well as information on population dynamics and their associated ecological factors. In this regard, the introduction into practice of self-organized networks of traps that collectively report data on local, regional, and even country scales using the Internet of Things (IoT) [31] could lead to a breakthrough in the adoption of IPM on a larger scale.

The Special Issue entitled "Integrated Pest Management in Horticulture" encompasses several of the above-mentioned topics, and the data and research presented should further encourage the implementation of pest management methods to reduce the application of synthetic plant protection products via a system approach.

Conflicts of Interest: The authors declare no conflict of interest.

References

1. Ehler, L.E. Integrated pest management (IPM): Definition, historical development and implementation, and the other IPM. *Pest Manag. Sci.* **2006**, *62*, 787–789. [CrossRef] [PubMed]
2. Lefebvre, M.; Langrell, S.R.H.; Gomez-y-Paloma, S. Incentives and policies for integrated pest management in Europe: A review. *Agron. Sustain. Dev.* **2015**, *35*, 27–45. [CrossRef]
3. Preti, M.; Verheggen, F.; Angeli, S. Insect pest monitoring with camera-equipped traps: Strengths and limitations. *J. Pest Sci.* **2021**, *94*, 203–217. [CrossRef]
4. Liu, C.; Zhai, Z.; Zhang, R.; Bai, J.; Zhang, M. Field pest monitoring and forecasting system for pest control. *Front. Plant Sci.* **2022**, *13*, 990965. [CrossRef]
5. Trematerra, P. Aspects related to decision support tools and Integrated Pest Management in food chains. *Food Control* **2013**, *34*, 733–742. [CrossRef]
6. Damos, P. Modular structure of web-based decision support systems for integrated pest management. A Review. *Agron. Sustain. Dev.* **2015**, *35*, 1347–1372. [CrossRef]

7. Skendžić, S.; Zovko, M.; Pajač Živković, I.; Lešić, V.; Lemić, D. Effect of climate change on introduced and native agricultural invasive insect pests in Europe. *Insects* **2021**, *12*, 985. [CrossRef] [PubMed]
8. Walsh, D.B.; Bolda, M.P.; Goodhue, R.E.; Dreves, A.J.; Lee, J.C.; Bruck, D.J.; Walton, V.M.; O'Neal, S.D.; Zalom, F.G. *Drosophila suzukii* (Diptera: Drosophilidae): Invasive pest of ripening soft fruit expanding its geographic range and damage potential. *J. Integr. Pest Manag.* **2011**, *1*, 1–7. [CrossRef]
9. Lee, D.H. Current status of research progress on the biology and management of *Halyomorpha halys* (Hemiptera: Pentatomidae) as an invasive species. *Appl. Entomol. Zool.* **2015**, *50*, 277–290. [CrossRef]
10. van Lenteren, J.C. The state of commercial augmentative biological control: Plenty of natural enemies, but a frustrating lack of uptake. *BioControl* **2012**, *57*, 1–20. [CrossRef]
11. Wajnberg, E.; Roitberg, B.D.; Boivin, G. Using optimality models to improve the efficacy of parasitoids in biological control programmes. *Entomol. Exp. Appl.* **2016**, *158*, 2–16. [CrossRef]
12. Bueno, A.D.F.; Sutil, W.P.; Maciel, R.M.A.; Roswadoski, L.; Colmenarez, Y.C.; Colombo, F.C. Challenges and opportunities of using egg parasitoids in FAW augmentative biological control in Brazil. *Biol. Control* **2023**, *186*, 105344. [CrossRef]
13. Raven, P.H.; Wagner, D.L. Agricultural intensification and climate change are rapidly decreasing insect biodiversity. *Proc. Natl. Acad. Sci. USA* **2021**, *118*, e2002548117. [CrossRef]
14. Lessando, M.; Gontijo, M. Engineering natural enemy shelters to enhance conservation biological control in field crops. *Biol. Control* **2019**, *130*, 155–163. [CrossRef]
15. Kleijn, D.; Biesmeijer, K.J.C.; Klaassen, R.H.G.; Oerlemans, N.; Raemakers, I.; Scheper, J.; Vet, L.E.M. Integrating biodiversity conservation in wider landscape management: Necessity, implementation and evaluation. In *Advances in Ecological Research*; Bohan, D.A., Vanbergen, A.J., Eds.; Academic Press: Cambridge, MA, USA, 2020; Volume 63, pp. 127–159. [CrossRef]
16. Gutierrez, A.P.; Caltagirone, L.E.; Meikle, W. Evaluation of results: Economics of biological control. In *Handbook of Biological Control*; Bellows, T.S., Fisher, T.W., Eds.; Academic Press: San Diego, CA, USA, 1999; pp. 243–252.
17. Hatt, S.; Boeraeve, F.; Artru, S.; Dufrene, M.; Francis, F. Spatial diversification of agroecosystems to enhance biological control and other regulating services: An agroecological perspective. *Sci. Total Environ.* **2018**, *621*, 600–611. [CrossRef]
18. Tscharntkea, T.; Karp, D.S.; Chaplin-Kramer, R.; Batary, P.; DeClerck, F.; Grattone, C.; Hunt, L.; Ives, A.; Jonsson, M.; Larsen, A.; et al. When natural habitat fails to enhance biological pest control—Five hypotheses. *Biol. Conserv.* **2016**, *204*, 449–458. [CrossRef]
19. Jonsson, M.; Wratten, S.D.; Landis, D.A.; Tompkins, J.-M.L.; Cullen, R. Habitat manipulation to mitigate the impacts of invasive arthropod pests. *Biol. Invasions* **2010**, *12*, 2933–2945. [CrossRef]
20. Tscharntke, T.; Bommarco, R.; Clough, Y.; Crist, T.O.; Kleijn, D.; Rand, T.; Tylianakis, J.; van Nouhuys, S.; Vidal, S. Conservation biological control and enemy diversity on a landscape scale. *Biol. Control* **2007**, *43*, 294–309. [CrossRef]
21. Borsotto, P.; Borri, I.; Tartanus, M.; Zikeli, S.; Lepp, B.; Kelderer, M.; Holtz, T.; Friedl, M.; Boutry, C.; Neri, D.; et al. Innovative agricultural management in organic orchards and perception of their potential ecosystem services. *Acta Hortic.* **2022**, *1354*, 1–8. [CrossRef]
22. Costanza, R.; D'Arge, R.; de Groot, R.; Farber, S.; Grasso, M.; Hannon, B.; Limburg, K.; Naeem, S.; O'Neill, R.V.; Paruelo, J.; et al. The value of the world's ecosystem services and natural capital. *Nature* **1997**, *387*, 253–260. [CrossRef]
23. Naranjo, S.E.; Ellsworth, P.C.; Frisvold, G.B. Economic value of biological control in Integrated Pest Management of managed plant systems. *Annu. Rev. Entomol.* **2015**, *60*, 621–645. [CrossRef] [PubMed]
24. Kevin, D.; Gallagher, K.D.; Ooi, P.A.C.; Kenmore, P.E. Impact of IPM Programs in Asian Agriculture. In *Integrated Pest Management: Dissemination and Impact*; Peshin, R., Dhawan, A.K., Eds.; Springer: Dordrecht, The Netherlands, 2009; pp. 347–358.
25. Freier, B.; Boller, E.F. Integrated Pest Management in Europe—History, Policy, Achievements and Implementation. In *Integrated Pest Management: Dissemination and Impact*; Peshin, R., Dhawan, A.K., Eds.; Springer: Dordrecht, The Netherlands, 2009; pp. 435–454. [CrossRef]
26. Jones, V.P.; Brunner, J.F.; Grove, G.G.; Petit, B.; Tangren, G.V.; Jones, W.E. A web-based decision support system to enhance IPM programs in Washington tree fruit. *Pest Manag. Sci.* **2010**, *66*, 587–595. [CrossRef] [PubMed]
27. Kuehne, G.; Llewellyn, R.; Pannell, D.J.; Wilkinson, R.; Dolling, P.; Ouzman, J.; Ewing, M. Predicting farmer uptake of new agricultural practices: A tool for research, extension and policy. *Agric. Syst.* **2017**, *156*, 115–125. [CrossRef]
28. Rose, D.C.; Sutherland, W.J.; Parker, C.; Lobley, M.; Winter, M.; Morris, C.; Twining, S.; Ffoulkes, C.; Amano, T.; Dicks, L.V. Decision support tools for agriculture: Towards effective design and delivery. *Agric. Syst.* **2016**, *149*, 165–174. [CrossRef]
29. Arcega Rustia, D.J.; Chiu, L.-Y.; Lu, C.-Y.; Wu, Y.-F.; Chen, S.-K.; Chung, J.-Y.; Hsu, J.-C.; Lin, T.-T. Towards intelligent and integrated pest management through an AIoT-based monitoring system. *Pest Manag. Sci.* **2022**, *78*, 4288–4302. [CrossRef] [PubMed]
30. Wandkar, S.; Bhatt, Y.C.; Jain, H.K.; Nalawade, S.M.; Pawar, S.G. Real-Time Variable Rate Spraying in Orchards and Vineyards: A Review. *J. Inst. Eng. Ser. A* **2018**, *99*, 385–390. [CrossRef]
31. Potamitis, I.; Eliopoulos, P.; Rigakis, I. Automated Remote Insect Surveillance at a Global Scale and the Internet of Things. *Robotics* **2017**, *6*, 19. [CrossRef]

Disclaimer/Publisher's Note: The statements, opinions and data contained in all publications are solely those of the individual author(s) and contributor(s) and not of MDPI and/or the editor(s). MDPI and/or the editor(s) disclaim responsibility for any injury to people or property resulting from any ideas, methods, instructions or products referred to in the content.

Article

Evaluation of *Artemisia absinthium* L. Essential Oil as a Potential Novel Prophylactic against the Asian Citrus Psyllid *Diaphorina citri* Kuwayama

Syed Arif Hussain Rizvi [1,*], Laila A. Al-Shuraym [2], Mariam S. Al-Ghamdi [3], Fahd Mohammed Abd Al Galil [4], Fahd A. Al-Mekhlafi [5], Mohamed Wadaan [5] and Waqar Jaleel [6,7]

[1] State Key Laboratory of Desert and Oasis Ecology, Xinjiang Institute of Ecology and Geography, Chinese Academy of Sciences, Urumqi 830011, China
[2] Department of Biology, College of Science, Princess Nourah bint Abdulrahman University, P.O. Box 88428, Riyadh 11671, Saudi Arabia; laalshuraym@pnu.edu.sa
[3] Department of Biology, Faculty of Applied Sciences, Umm Al-Qura University, Makkah 24381, Saudi Arabia; msghamidy@uqu.edu.sa
[4] Department of Biology, Faculty of Science, University of Bisha, P.O. Box 551, Bisha 61922, Saudi Arabia; fahd.bamu78@gmail.com
[5] Department of Zoology, College of Science, King Saud University, P.O. Box 2455, Riyadh 11451, Saudi Arabia; falmekhlafi@ksu.edu.sa (F.A.A.-M.); wadaan@ksu.edu.sa (M.W.)
[6] Entomology Section, Central Cotton Research Institute, Multan P.O. Box 66000, Pakistan; waqarjaleel4me@yahoo.com
[7] Entomology, Horticultural Research Station, Bahawalpur P.O. Box 63100, Pakistan
* Correspondence: arifrizv@aup.edu.pk or rizvi@ms.xjb.ac.cn

Simple Summary: The extensive use of synthetic chemical pesticides to manage insect pests has adversely affected humans and the environment. Therefore, botanical pesticides could be helpful as alternative tools for integrated pest management since they have low mammalian toxicity and minimal risk of developing resistance in target pests. Our current study aimed to evaluate the efficacy of *Artemisia absinthium* essential oil (AAEO) as a novel alternative to synthetic insecticides against *Diaphorina citri*. The results indicated that AAEO could be developed as a valuable novel crop protectant against *D. citri*.

Abstract: Interest in developing novel crop protectants has increased in the recent decade due to the harmful effects of synthetic pesticides on humans and the environment. *Diaphorina citri* threatens the citrus industry worldwide and is the primary vector of phloem-limited bacterium (HLB). However, there is no available cure for HLB. *Diaphorina citri* management mainly depends on the use of synthetic insecticides, but their massive application leads to resistance in pest populations. Therefore, alternative pest management strategies are needed. Our results indicated that fewer *D. citri* adults settled on plants treated with AAEO than on control 48 h after release. The psyllids fed on citrus leaves treated with AAEO significantly reduced the honeydew production compared to the control. The AAEO showed potent ovicidal activity against the *D. citri* eggs with LC_{50} 5.88 mg/mL. Furthermore, we also explored the fitness of *D. citri* on AAEO-treated and untreated *Citrus sinensis* by using two-sex life table tools. Our results revealed that the intrinsic rate of increase (r) was higher on untreated seedlings (0.10 d^{-1}) than those treated with an LC_{20} concentration of AAEO (0.07 d^{-1}). Similarly, the net reproductive rate (R_0) was higher for untreated seedlings (14.21 offspring) than those treated (6.405 offspring). Furthermore, the AAEO were safer against *Aphis mellifera*, with LC_{50} 35.05 mg/mL, which is relatively higher than the LC_{50} 24.40 mg/mL values against *D. citri*. The results indicate that AAEO exhibits toxic and behavioral effects on *D. citri*, which can be a potential candidate for managing this pest.

Keywords: essential oil; toxicity; botanical insecticides; *Artemisia absinthium*; *Diaphorina citri*

1. Introduction

The *Diaphorina citri* Kuwayama (Hemiptera: Psyllidae) is a well-known significant citrus pest across the globe. Huanglongbing (HLB) is the most devastating disease of *Citrus* spp. worldwide, limiting commercial production [1]. The HLB-infected trees show abnormal growth, yellowing of leaves, mottling of leaves, and lack of fruit juice. Eventually, the whole plant may die [2,3]. However, no cure for HLB has been reported [4] and controlling *D. citri* is one of the effective ways to manage HLB [5]. Currently, the primary control measures of *D. citri* heavily rely on synthetic insecticides [6,7].

Since 2005, the citrus industry in China, Brazil, and Florida has been massively devastated by HLB, which caused a loss of 74% of production [8]. The management of *D. citri* has played a significant role in HLB dispersal in citrus groves. Eight to twelve insecticide applications are commonly practiced in one cropping season in major citrus-growing countries worldwide [9]. However, the massive and repetitive use of these synthetic insecticides results in environmental pollution and the development of resistance [9,10]. Therefore, there is a need to develop novel eco-friendly alternatives for managing *D. citri*.

Since ancient times, botanical pesticides have been used to manage various agricultural and household pests, including *Azadirachta indica* and *Nicotiana tabacum* [11,12]. Botanical pesticides are safe because they cause no or minimal toxicity to humans and mammals [13]. Furthermore, these pesticides are biodegradable and do not produce any toxic chemicals in the environment [14].

In the recent decade, essential oils (EOs) have been gaining attention among the scientific community as novel alternatives to synthetic chemicals due to their rapid biodegradability and ability to disrupt insect biochemical, physiological, and behavioral functions [15,16]. *Artemisia absinthium* L. is an aromatic and medicinal herb and it has been reported to have toxicity against various agricultural and household pests, including store pests *Callosobruchus maculatus* and *Bruchus rufimanus* [17], repellent and larvicidal activity against *Aedes aegypti* and *Musca domestica* [18], contact and residual toxicity against *Sitophilus oryzae* [19], and fumigant mode of action against *Solenopsis geminata* and *Tribolium castaneum* [20].

Therefore, our study aimed to evaluate the *A. absinthium* (AAEO) toxicity and behavioral effect against *D. citri*. Furthermore, the effect of AAEO on the fitness of *D. citri* was also studied using two-sex life table tools.

2. Materials and Methods

2.1. Insects

The *D. citri* adults were collected from *Murraya paniculata* L., grown at the South China Agricultural University Guangzhou, China. The psyllids were allowed to acclimatize to laboratory conditions (28 ± 2 °C, 65 ± 5% relative humidity (RH), photoperiod 14:10 (Light: Dark)). Males and females were separated based on their morphological characteristics. The yellow or orange color of the female abdomen indicates that it contains eggs [21,22].

2.2. Plant Materials Extraction Procedure

The aerial parts of *A. absinthium* were collected from Skardu Baltistan, Pakistan, in August 2022. Our previously published paper reported the plant species and its GC-MS analysis [23] (Table 1). The EOs were stored in transparent glass vials and kept at 4 °C for further experimentation.

Table 1. AAEO dominant constituents were identified through GC/MS [23].

Peak #	Compounds Name [b]	Relative %	RT [a]	KI (Exp) [c]
1	β-myrcene	0.86	12.351	1147
2	Pinocarvone	0.62	13.291	1172
3	α-Gurjunene	1.68	21.838	1416

Table 1. *Cont.*

Peak #	Compounds Name [b]	Relative %	RT [a]	KI (Exp) [c]
4	α-Humulene	0.94	22.976	1452
5	α-Copaene	3.51	23.811	1478
6	g-Curcumene	0.45	24.054	1486
7	*epi*-Cubenol	2.67	24.754	1508
8	β-Calacorene	2.10	26.599	1570
9	(-)-Spathulenol	1.94	26.737	1575
10	Germacrene-D-4-ol	3.48	26.861	1579
11	Guaiol	19.34	27.559	1602
12	Thujol	2.69	27.837	1620
13	4-*epi*-Cubedol	1.68	28.356	1631
14	Cubenol	1.89	28.64	1641
15	γ-Eudesmol	1.19	29.011	1654
16	8-*epi*-γ-Eudesmol	1.14	29.113	1657
17	a-Cadinol	2.76	29.269	1663
18	Geranial	8.83	29.844	1686
19	Chamazulene	5.94	31.067	1728
20	1,3-Dicyclopentylcyclopentane	0.93	31.455	1746
21	Fraganol	0.95	32.355	1769
22	Tetrakis(1-methyl)-Pyrazine	2.26	32.92	1797
24	Cubedol	1.16	36.568	1941
25	Geranyl-*p*-Cymene	1.63	36.748	1948
26	Nerolidol-epoxyacetate	1.12	37.999	1999
27	Geranyl-α-terpinene	5.64	38.176	2007
28	Spathulenol	0.83	39.549	2066
29	Heneicosane	1.60	40.341	2100
30	Eugenol	1.21	40.507	2102
31	Carvacrol	5.47	41.557	2147
32	α-Bisabolol	6.17	41.721	2166
33	1-ethyl-4-methoxy-benzene	0.53	43.784	2256
34	Tricosane	1.48	44.735	2300
35	1-Heptatriacotanol	1.03	44.931	2309
36	Pentacosane	2.20	48.786	2500
37	Heptacosane	1.28	52.539	2700
38	Nonacosane	0.80	56.106	2899
	Total identified	99.9		
	Oil yield (%)	0.46		
	Monoterpenes	20.42		
	Sesquiterpenes	52.69		
	Others	26.89		

[a] Retention time. [b] Compounds are listed in order of their retention time. [c] Retention index relative to C_7-C_{40} *n*-alkanes on a DB-1 (30 m × 0.22 mm i.d., 0.25 μm film thickness). Identification methods: RI, based on comparison of calculated RI with those reported in Adams or NIST 08 and previous literature.

2.3. Settling Behaviour of D. citri

The attractiveness and settling behavior of *D. citri* adults towards sweet orange seedlings (10–15 cm in length) treated with desired concentrations 0.1, 0.5, 1, 2, 3% w/v of AAEO, diluted in 20% ethanol containing 0.01% Tween 80, which corresponds to the dosage of 1, 5, 10, 15, 20 and 30 mg/mL, respectively, were observed in a choice experiment under controlled laboratory conditions (27 ± 2 °C, 65 ± 5% RH, photoperiod 14:10 (Light: Dark). One hundred adults (50 male and 50 female) were released into the center of cages, each cage with six *C. sinensis* seedlings. Each seedling was sprayed using a mini plastic trigger sprayer (Deqing Yuanchen Plastic Products Co., Ltd., Deqing, China) with 1 mL of the desired concentrations of AAEO and was allowed to dry under the hood. The flasks were randomly positioned inside the cage. There was a total of five replicate cages. The total numbers of *D. citri* adults settling on each seedling were recorded after an interval of 12 and 24 h after release. Within 2 h after release, the cages were examined to check the mortality due to mechanical injury while aspirating was discarded. The guava crude methanolic extract at 30 mg/mL concentration was used as a positive control, as much literature indicates that Guava repels *D. citri* [24,25]. The number of *D. citri* adults that settled on each seedling under the various treatments was compared.

2.4. Antifeedant Activity of AAEO against D. citri

The amount of honeydew excreted by the adults while they were kept to *C. sinensis* seedlings treated with different concentrations of AAEO was used to assess the feeding activity of *D. citri*. The feeding bioassay arenas comprised 1.5% agar solution-coated mini glass Petri dishes [26]. Freshly excised leaves from *C. sinensis* were used for all bioassays. The leaf disk, average size 5.50 ± 0.3 cm in length, was dipped for 5 s in the desired concentrations of AAEO and allowed to dry for one hour under the fume hood. Ten CO_2 anaesthetized *D. citri* adults were released in each Petri dish, and the Petri dishes were capped with a lid lined with 60 mm filter paper. One hour after release, the Petri dish was examined to check the mortality due to mechanical injury while aspirating was discarded. Then, the Petri dishes were closed with lab parafilm and turned upside down. The filter papers were removed and immersed in 1% w/v ninhydrin 48 h after release for three minutes and were then dried at room temperature [27]. The feeding activities of *D. citri* were measured by calculating the number of purple spots.

2.5. Ovicidal Toxicity of AAEO

Ovicidal activities AAEO against *D. citri* were evaluated by confining the eggs containing sprayed *C. sinensis* seedlings with an aerial insect net (25 cm length, 20 cm width). Sixteen mixed populations (eight male and eight female) of *D. citri* adults were aspirated. The psyllids were released on the potted seedlings to lay eggs, while the pots and seedlings were covered with an insect net. The psyllids were removed from the gauze nets five days after release, and the number of eggs laid was calculated. The seedlings with eggs were treated with desired concentrations of 0.1, 0.5, 1, 2, and 3% w/v of AAEO. Then, the seedlings were confined with gauze nets and the number of hatched nymphs was counted until all the eggs were hatched. The ovicidal activity was assessed regarding egg mortality rate (EMR) using the formula below.

$$\text{EMR (\%)} = \frac{Number\ of\ eggs\ unhatched}{Total\ number\ of\ eggs\ laid} \times 100.$$

2.6. Effect of AAEO on Fitness and Development of D. citri

The laboratory conditions were equal to those during the ovicidal toxicity bioassay. Briefly, the *C. sinensis* seedlings were sprayed with an LC_{20} concentration (8.3 mg/mL) of AAEO. After drying, *D. citri* virgin adults (five male and five female) were released in the bioassay arena for mating and oviposition. Five days after release, the adults were removed, and the number of eggs laid in each cage was calculated. The seedlings were observed daily until adult emergence was completed as well as the data of development time from eggs to adult formation, after adult formation as pre-oviposition. The following equations [28,29] were used to compute the population growth rate (*PGR*), oviposition, and fecundity using the age stage two-sex life table software.

$$PGR = \frac{(Nf/No)}{\Delta t},$$

whereas

N_f = Final number of *D. citri*;
N_0 = Initial number of *D. citri*;
Δt = Total number of days for the experiment.

The result with positive values indicated an increasing population, PGR = 0 indicated a stable population, while negative values indicated a decline in population and led towards extinction.

2.7. Toxicity of AAEO against Non-Targeted Organisms

To evaluate the toxicity of AAEO against *Apis mellifera*, no-choice feeding bioassays were used [30] under laboratory conditions in a plastic container (0.5 L) by following the procedure described previously [31]. Briefly, the following concentrations of AAEO (24, 36, 48, 60, and 72 mg/mL) were prepared in a 50% sugar solution. Ten healthy foraging workers were introduced into the cage. Each concentration was repeated thrice, and the control contained only 50% sugar solution. The mortality data were recorded within 48 h after treatment. Each container was considered a single treatment and each treatment was replicated five times. Five replications of each container were used to represent one treatment. Imidacloprid was used as a positive control at the following concentrations (5, 10, 15, 20 and 25 ug/mL).

2.8. Statistical Analysis

Chi-square goodness of fit tests were used to estimate the significance of choice between treated and untreated seedlings. The toxicity data were assessed by using Probit analysis (SPPSS 17.0). According to Levene's test, all data sets were homoscedastic and the mean difference between treatments was separated by using Tukey's HSD test. The population parameters of *D. citri* were estimated using the age stage two-sex life table program. The population and age stage parameters of *D. citri*, e.g., R_0, r, k and $s_{xj}, f_{xj}, l_x, m_x, e_{xj}, v_{xj}$, respectively, were calculated as described in the methodology [32]. The two-sex life table was calculated using the following formulae:

$$l_x = \sum_{j=1}^{k} S_{xj} \quad (1)$$

$$m_x = \frac{\sum_{j=1}^{k} S_{xj} f_{xj}}{\sum_{j=1}^{k} S_{xj}} \quad (2)$$

$$R_0 = \sum_{x=0}^{\infty} l_x m_x, \quad (3)$$

where k is the number of stages. This study used the Euler–Lotka formula's iterative bisection approach to estimate the r, with the age index starting at 0, as in Equation (2).

$$\sum_{x=0}^{\infty} e^{-r(x+1)} l_x m_x = 1 \quad (4)$$

$$e_{xj} = \sum_{i=x}^{\infty} \sum_{y=j}^{k} s'_{iy} \quad (5)$$

$$V_{xj} = \frac{e^{r(x=1)}}{S_{xj}} \sum_{i=x}^{\infty} e^{-r(x+1)} \sum_{y=j}^{k} s'_{iy} f_{iy}. \quad (6)$$

3. Results

3.1. Effect of AAEO on Settling Behavior of D. citri

Overall, concentration and time-dependent effects were observed in the settling behavior of *D. citri* adults. The settling behavior of *D. citri* adults was not significantly different among the various AAEO concentrations tested compared with the control at 24 h (F = 18.98; df = 4, 24; p = 0.243) after release. However, significant differences were observed after 48 h (F = 66; df = 4, 24; p = 0.005) and 72 h (F = 86; df = 4, 24; p = 0.001) after release (Figure 1). On control plants, compared to those treated with AAEO, more *D. citri* adults were recorded after 72 h.

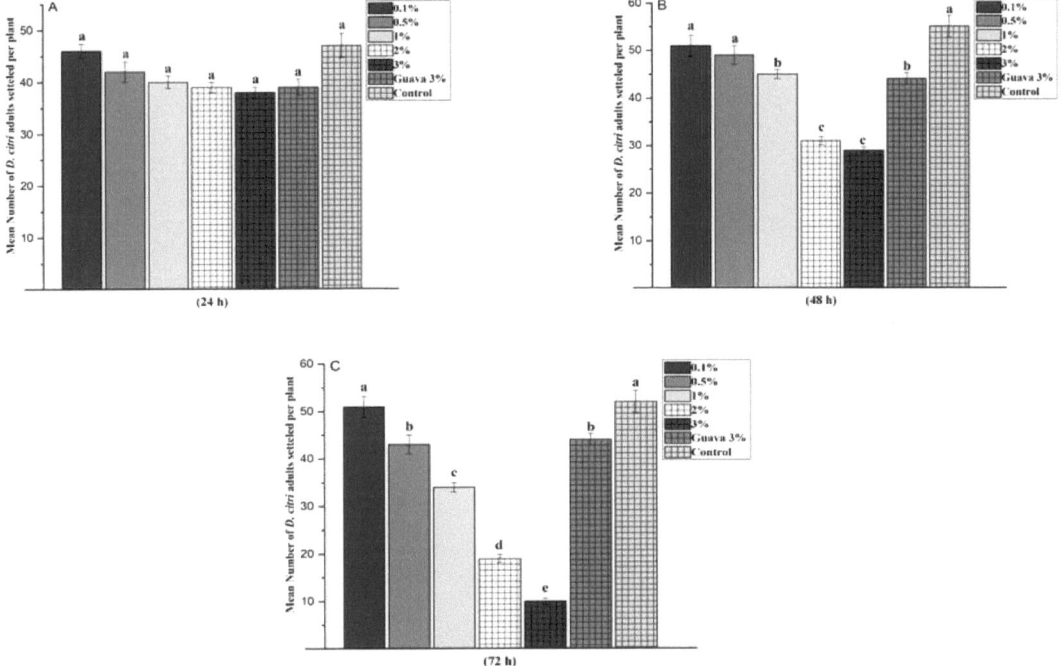

Figure 1. Settling preference of *D. citri* adults on seedlings treated with various concentrations of AAEO 24 (**A**), 48 (**B**), and 72 h (**C**) after the release of adults. Bars within a panel not labelled by the same letter are significantly different according to Tukey's test ($p < 0.05$).

3.2. Effect of AAEO on D. citri Feeding Activity

The feeding activity of *D. citri* measured by the amount of honeydew extraction was presented in (Figure 2). The results indicated a concentration-dependent antifeedant effect of AAEO on the feeding activity of *D. citri*. Except for 1 mg/mL of AAEO, all the treatments, including 5, 10, 20, and 30 mg/mL, reduced the excretion of honeydew significantly in comparison to the control ($F = 84.47$; $df = 4, 24$; $p < 0.0001$). However, there was a 92 and 86% honeydew excretion reduction by AAEO at 20 and 30 mg/mL.

3.3. Effect of AAEO on Eggs Hatchability of D. citri

The results indicated that a concentration-dependent response of AAEO on the egg hatchability of *D. citri* was observed. The AAEO has shown potent ovicidal activity with an LC_{50} 5.88 mg/mL. The number of eggs hatchability per plant was significantly lower than for the control, except for 1 mg/mL AAEO ($F = 63.82$; $df = 5, 29$; $p < 0.0023$). On sweet orange potted plants treated with 30 mg/mL AAEO, only 11.75% of eggs were able to hatch into adults, followed by 5, 10, and 20 mg/mL, where only 30.44, 72.46, and 83.65% of eggs were able to hatch into adults, respectively.

3.4. Population Parameters

The effect of AAEO on the population parameters of *D. citri* indicated that there was an increase in the intrinsic rate (r), which was higher on untreated sweet oranges (0.10 d^{-1}) than on those treated with an LC_{20} concentration of AAEO (0.07 d^{-1}). Similarly, the net reproductive rate (R_0) was higher for untreated *C. sinensis* seedlings (14.21 offspring) than for those treated (6.40 offspring) with LC_{20} concentration of AAEO (Table 2).

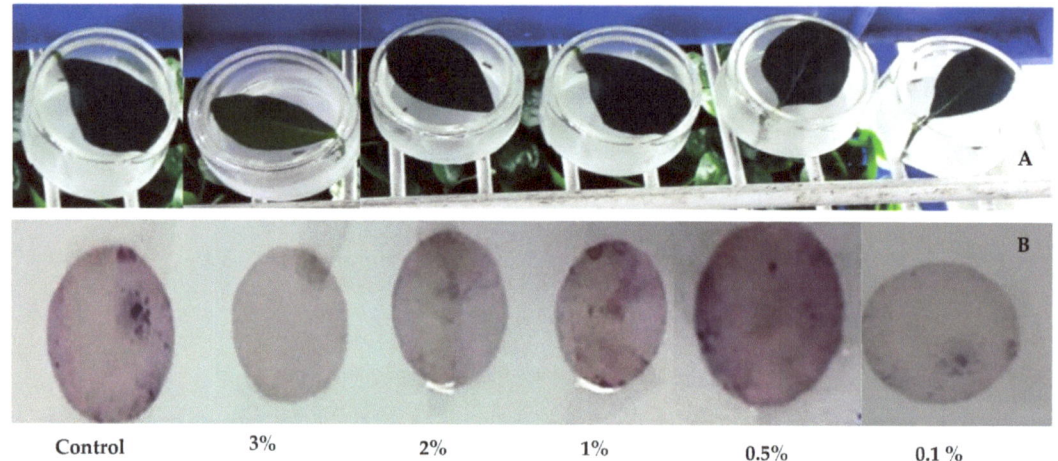

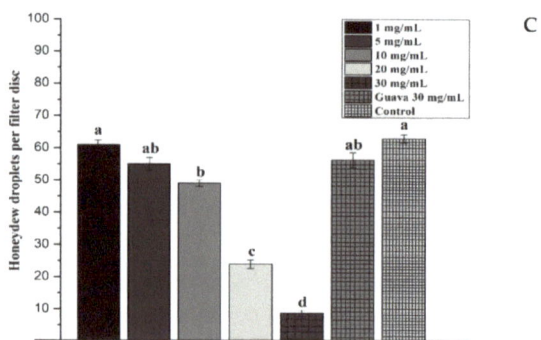

Figure 2. Effect of AAEO on *D. citri* adult feeding as measured by the number of honeydew droplets produced. (**A**) Citrus leaf discs treated with various concentrations of AAEO in 20% ethanol + 0.05% Tween 80 or 20% ethanol + 0.05% Tween 80 (as control) were exposed to 10 *D. citri* adults, (**B**) The filter papers with honeydew droplets produced by *D. citri* adults, filter paper immersed in 1% ninhydrin forms a purple spot. (**C**) Effect of AAEO on *D. citri* adult feeding. Bars within a panel not labelled by the same letter are significantly different according to Tukey's test ($p < 0.05$).

Table 2. Effect of AAEO on reproductive and population parameters of the *D. citri* on treated sweet orange seedlings compared to untreated sweet orange seedlings.

Traits	Treated	Untreated
r (per day)	0.07	0.10
λ (per day)	1.07	1.11
GRR (offspring)	7.53	16.02
R_0 (offspring/individual)	6.40	14.21

r; The intrinsic rate of increase (per day) λ; The finite rate of increase (per day) GRR; Gross reproductive rate (offspring) R_0; The net reproductive rate (offspring/individual).

Diaphorina citri, the comprehensive age-stage survival rate (sxj) on treated and untreated *C. sinensis* seedlings, was determined. Our findings showed the possibility of a freshly hatched larva making it to age *x* and stage *j* (Figure 3) because development rates varied across individuals on treated and untreated seedlings; similarly, the projected curves exhibited completely different layouts at each developmental stage. Individual survival rates rapidly dropped with age and showed an inverse relationship between treated and untreated seedlings (Figure 3). The developmental time of *D. citri* females was long, and the survival rate was shorter on the untreated than the treated ones, while in the case of males, the development was shorter, and the survival rate was longer on untreated compared to on treated *C. sinensis* seedlings (Figure 3).

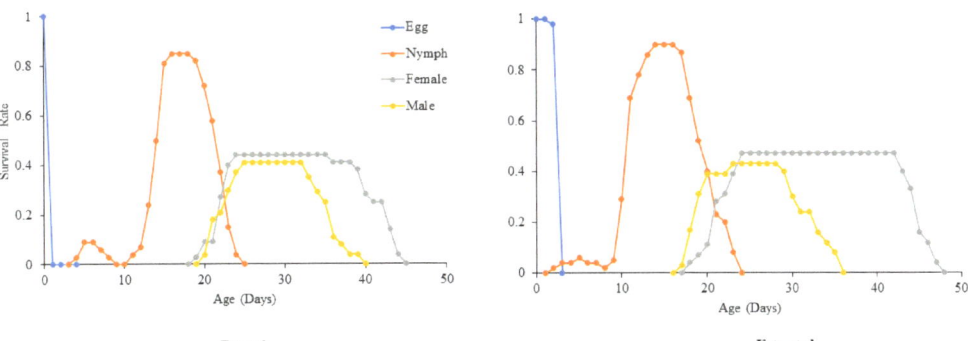

Figure 3. Effect of AAEO on the age-stage-specific survival rate (s_{xj}) of the *D. citri* on sweet orange seedlings compared to untreated sweet orange seedlings.

The highest value of age-stage specific fecundity (f_{xj}) was higher on untreated sweet oranges than on treated ones (Figure 3). There is a direct relation to the age-specific maternity (l_x*m_x) of *D. citri*. As the survival rate increases, fecundity increases in treated and untreated cases. However, the constant peak point of age-specific maternity (l_x*m_x) of *D. citri* was higher on untreated sweet oranges than on *C. sinensis* seedlings treated with AAEO (Figure 4).

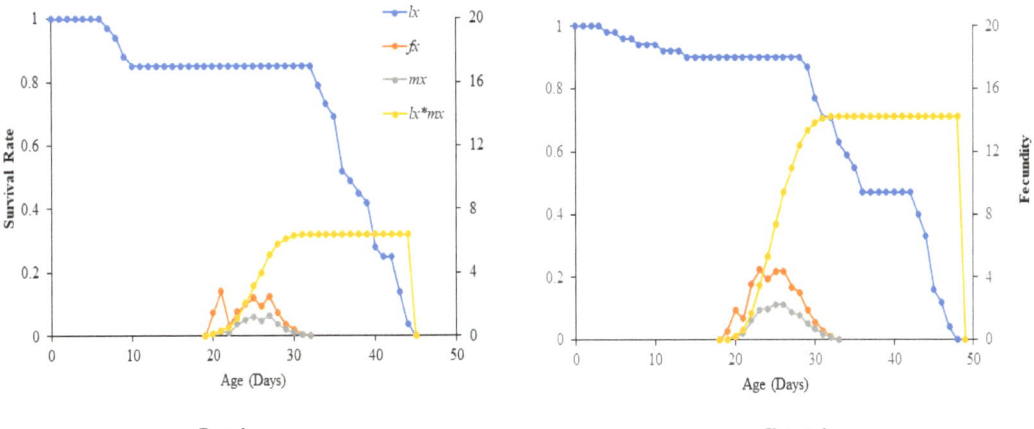

Figure 4. Effect of AAEO on the age-specific survival rate (l_x), female age-specific fecundity (f_x), age-specific fecundity (m_x), and age-specific maternity (l_x*m_x) of the *D. citri* on sweet orange seedlings in comparison to untreated sweet orange seedlings.

The effects of treated and untreated sweet oranges on the population's expected average life expectancy (e_{xj}) at egg, nymph, and adult stages of *D. citri* were determined in (Figure 5). The longevity of the newly hatched *D. citri* eggs was longer in untreated sweet oranges than in treated ones. The peak life expectancy (e_{xj}) value for female adults of *D. citri* was higher on untreated sweet oranges than on treated oranges (Figure 5). The *exj* of male adults of *D. citri* was the maximum on treated sweet oranges compared to untreated oranges. Overall, all stages of the highest life expectancy (e_{xj}) of *D. citri* were recorded on untreated sweet oranges (Figure 5). The age-stage reproductive value (v_{xj}) of *D. citri* in (Figure 6) explains an individual's role in the future population (i.e., the population forecasting scale) at age x and stage j.

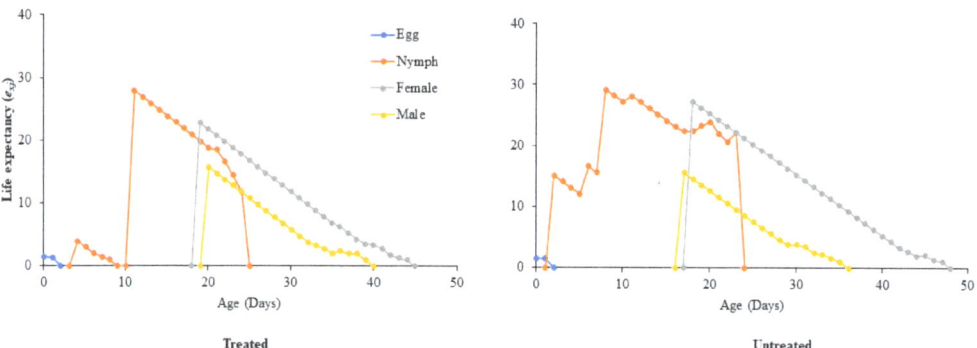

Figure 5. Effect of AAEO on the life expectancy (e_{xj}) of the *D. citri* on sweet orange seedlings compared to untreated sweet orange seedlings.

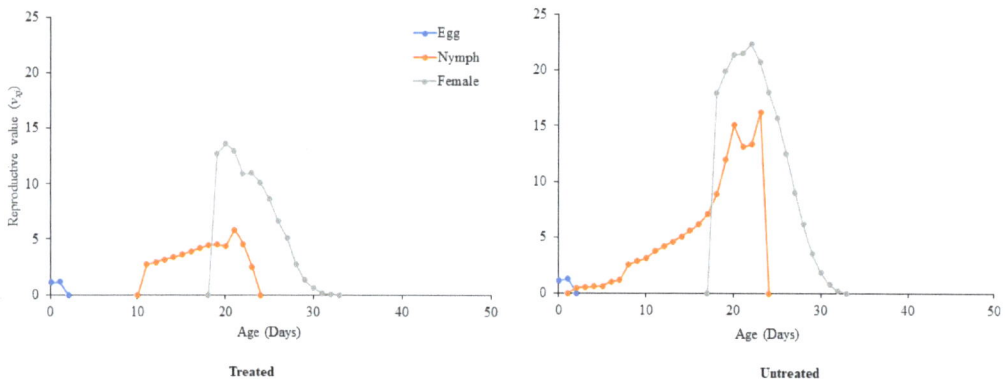

Figure 6. Effect of AAEO on the reproductive value (v_{xj}) of the *D. citri* on sweet orange seedlings compared to untreated sweet orange seedlings.

3.5. Toxicity of AAEO against A. mellifera

The AAEO caused toxicity against *A. mellifera* at significantly higher concentrations, with LC_{50} and LC_{90} of 35.05 and 55.86 mg/mL, respectively, which is too high compared to the lethal dose of AAEO against *D. citri* LD_{50} of 5.20 µg/insect via the topical application method recorded in our previous published paper [33] (Table 3). Therefore, AAEO can be considered safe against *A. mellifera*.

Table 3. Toxicity of AAEO against *A. mellifera*.

Concentration (mg/mL)	Exposed [a]	% Mortality ± SD [b]	LC_{50} [c] (95% CL [d])	LC_{90} [e] (95% CL)	X^2 (df) [f]	*p*-Value
24	64	21.65 ± 0.87	35.05 (25.58–34.69)	55.86 (46.01–67.81)	0.94 (2)	0.23
36	60	35.32 ± 0.59				
48	67	52.01 ± 0.87				

Table 3. Cont.

Concentration (mg/mL)	Exposed [a]	% Mortality ± SD [b]	LC$_{50}$ [c] (95% CL [d])	LC$_{90}$ [e] (95% CL)	X^2 (df) [f]	p-Value
60	67	65.34 ± 0.60				
72	63	84.32 ± 0.51				
Control	65	4.11 ± 0.11				
Imidacloprid (ug/mL)						
5	66	20.22 ± 0.23	13.49 (9.45–19.34)	47.87 (33.32–55.31)	4.41 (3)	0.49
10	58	29.11 ± 0.35				
15	69	48.04 ± 0.45				
20	61	65.21 ± 0.32				
25	56	81.00 ± 0.13				
Control	62	7.12 ± 0.99				

[a] Total number of bees treated. [b] SD. Standard deviation. [c] LC$_{50}$. 50 % lethal concentration. [d] CL. Confidence limits. [e] LC$_{90}$. 90% lethal concentration. [f] χ^2 chi-square, df degrees of freedom.

4. Discussion

Botanical pesticides are plant-based substances, including pyrethrin, azadirachtin, neem, garlic, and vegetable oil. Because they degrade quickly in the presence of light and air, botanicals often have a short shelf life in the environment. [34]. These include plant extracts and essential oils, which are eco-friendly, biodegradable, and nontoxic to mammals [33]. The essential oils from various plants are reported to have antifeedant, repellent, and toxic activities against many insect pests [35]. Eos are complex mixtures of different compounds but are majorly dominated by monoterpenes and sesquiterpenes [36]. These monoterpenes and sesquiterpenes exert different toxic and behavioral effects against insects. For example, limonene decreased oviposition in mite *Oligonychus ununguis* [37], carvacrol and thymol showed contact toxicity against *Pochazia shantungensis* [38], α-pinene, eucalyptol and camphor exerted both fumigant and antifeedant activities against *Solenopsis invicta* and *Meloidogyne incognita* [39], thymol showed substantial contact toxicity against *Blattella germanica* (Yeom et al., 2012), and 1,8-cineole showed a fumigant mode of action against *Solenopsis invicta* and *Ectomyelois ceratoniae* [40,41]. Similarly, carvacrol showed contact toxicity against *D. citri* [33].

Plant volatiles are crucial in herbivore host location and recognition [42]. Odors and plant colors mediate how herbivore insects find and recognize their potential host [43]. The *D. citri* relies on its olfaction to locate and evaluate its potential host [44]. The volatile chemicals released by non-host plants obscure the host plant odor that phytophagous insects detect, leading to host plant avoidance and non-preference [45]. Regarding the repellent activity of AAEO against *D. citri*, results indicated a concentration-dependent effect. The adults strongly preferred settling on the control *C. sinensis* seedlings to the treated seedlings. Compared to the control, the adult *D. citri* settling did not significantly decrease 24 h after release. However, only a few adults were seen on the treated plant 48 and 72 h after release compared to the control because it took *D. citri* just around 9 h to distinguish volatiles from host plants, from a combination of volatiles from non-hosts in the open atmosphere [46]. A literature report indicated that the host finding and recognition ability of *D. citri* were reduced when non-host plant semiochemicals were used [45,47]. Many non-host plants have shown repellent activities against *D. citri*, including Guava [3,24,48], *Allium* spp [49,50]. HLB bacteria can only multiply in the body of the eukaryotic host [51]. The transmission of HLB bacterium from the infected to the uninfected tree was primarily taken by nymphs and adults of *D. citri* [52]. Here, we found that AAEO reduces the feeding activity of *D. citri*, measured as the number of purple spots on the treated leaf disc. AAEO at 20 and 30 mg/mL reduced honeydew secretion by 72.86 and 85.5%, respectively. However, the effect of AAEO on *D. citri* feeding in terms of the number of honeydew droplets recorded per filter paper disc was lower than cyantraniliprole, a synthetic anthranilic diamide insecticide, which caused an 80% reduction in honeydew droplets secretion by

D. citri at 0.1 µg/mL [53]. To better understand the antifeedant activity of AAEO against D. citri, further investigation should be conducted using electrical penetration graph (EPG) technology.

Essential oils (EOs) are effective against several insect species. They act as growth inhibitors, toxins, deterrents, repellents, and toxicants [54]. The EOs of azadirachtin and *Piper aduncum* against nymph and adults of *D. citri* caused 90–100% mortality in nymph and below 80% in adults, being nontoxic to ectoparasitoid *Tamarixia radiata* (Hymenoptera: Eulophidae) [26]. Similarly, *Syzygium aromaticum*, *Eucalyptus obliqua*, *Tithonia diversifolia*, and *Citrus limonia*l EOs showed considerable toxicity and repellent effects against *D. citri* [55,56]. Primarily *D. citri* management was prodigiously focused on controlling adult psyllids. For this, many classes of insecticides have been utilized [57]. Limited literature is available regarding the developmental and ovicidal products against *D. citri*. According to the current study, AAEO-treated potted *C. sinensis* seedlings exhibit concentration-dependent ovicidal action. When enclosed with the dry residue of AAEO at 20 and 30 mg/mL, respectively, only 11.75 and 30.44 eggs were able to hatch into adults, whereas in the control and 10 mg/mL concentrations, the hatching percentages were 93.45 and 93.78, respectively. The result showed that AAEO has ovicidal activity against *D. citri*. However, AAEO ovicidal activity was lower than that of cyantraniliprole, a synthetic anthranilic diamide insecticide, which caused complete inhibition of *D. citri* eggs' hatchability at 0.025 µg/mL [53].

The potential mechanism of action of EOs against insects is neurotoxicity [58,59]. These EOs typically contain complex combinations of sesquiterpenes, biogenetically related phenols, and monoterpenes. These compounds have a variety of hydrophilic and hydrophobic properties that can easily permeate insect cuticles and disrupt their physiological processes [60,61]. Despite the most promising properties of EOs as a natural insecticide, many technical issues are raised for their broader application due to their rapid volatility and poor water solubility [15]. There are many challenges and constraints related to the commercialization of EOs, including strict legislation and lack of raw material availability [60]. Their rapid degradability and low persistence may significantly reduce toxicity [60], transient effects, and a dearth of high-quality raw materials that are affordably priced [61]. Due to their rapid breakdown and short persistence, EOs may have significantly lower toxicity [61]. However, the efficacy and persistence of EOs can be enhanced by encapsulation, nanoparticles, and nano gel formulation, and cyclodextrins [62]. Overall, EOs have the potential to develop eco-friendly candidates for novel pest management, which should be a top priority for preserving ecosystems from contamination. EOs and plant extracts are safer for the environment, humans, and non-targeted organisms than synthetic insecticides [63]. The AAEO caused toxicity against *A. mellifera* at significantly higher concentrations, with an LC_{50} value of 35.05 mg/mL, which is too high compared to the LD50 value of 5.20 µg/insect of AAEO against *D. citri*.

5. Conclusions

The current study found that AAEO had insignificant toxicity toward honeybees when evaluated as non-targeted organisms and demonstrated repellent and ovicidal properties against adult *D. citri*. More research should be pursued regarding its broader applicability and effect on natural enemies. It was concluded that the AAEO might be developed as a novel prophylactic against *D. citri* with the edge of being environmentally friendly.

Author Contributions: Conceptualization, S.A.H.R. and L.A.A.-S. methodology S.A.H.R. and L.A.A.-S.; software, W.J.; formal analysis, M.S.A.-G.; resources, S.A.H.R.; data curation, F.M.A.A.G. and W.J.; initial draft preparation, S.A.H.R.; review and editing, S.A.H.R., F.M.A.A.G. and M.W.; visualization, M.S.A.-G. and F.A.A.-M.; supervision, F.A.A.-M. and L.A.A.-S.; project administration, funding acquisition F.A.A.-M. and L.A.A.-S. All authors have read and agreed to the published version of the manuscript.

Funding: Project number (PNURSP2023R365), Princess Nourah bint Abdulrahman University, Riyadh, Saudi Arabia. Project (RSP2023R112), King Saud University, Riyadh, Saudi Arabia.

Institutional Review Board Statement: Not applicable.

Informed Consent Statement: Not applicable.

Data Availability Statement: Data are contained within the article.

Acknowledgments: The authors thank Princess Nourah bint Abdulrahman University Researchers Supporting Project number (PNURSP2023R365), Princess Nourah bint Abdulrahman University, Riyadh, Saudi Arabia. We also thank the Researchers Supporting Project (RSP2023R112), King Saud University, Riyadh, Saudi Arabia.

Conflicts of Interest: The authors declare no conflict of interest.

References

1. Bové, J.M. Huanglongbing: A destructive, newly-emerging, century-old disease of citrus. *J. Plant Pathol.* **2006**, *88*, 7–37.
2. Rizvi, S.A.H.; Ling, S.; Tian, F.; Liu, J.; Zeng, X. Interference mechanism of *Sophora alopecuroides* L. alkaloids extract on host finding and selection of the Asian citrus psyllid *Diaphorina citri* Kuwayama (Hemiptera: Psyllidae). *Environ. Sci. Pollut. Res. Int.* **2019**, *26*, 1548–1557. [CrossRef] [PubMed]
3. Zaka, S.M.; Zeng, X.N.; Holford, P.; Beattie, G.A.C. Repellent effect of guava leaf volatiles on settlement of adults of citrus psylla, *Diaphorina citri* Kuwayama, on citrus. *Insect Sci.* **2010**, *17*, 39–45. [CrossRef]
4. Halbert, S.E.; Manjunath, K.L. Asian citrus psyllids (Sternorrhyncha: Psyllidae) and greening disease of citrus: A literature review and assessment of risk in Florida. *Fla. Entomol.* **2004**, *87*, 330–353. [CrossRef]
5. Byrne, F.J.; Daugherty, M.P.; Grafton-Cardwell, E.E.; Bethke, J.A.; Morse, J.G. Evaluation of systemic neonicotinoid insecticides for the management of the Asian citrus psyllid *Diaphorina citri* on containerized citrus. *Pest Manag. Sci.* **2017**, *73*, 506–514. [CrossRef]
6. Srinivasan, R.; Hoy, M.A.; Singh, R.; Rogers, M.E. Laboratory and field evaluations of Silwet L-77 and kinetic alone and in combination with imidacloprid and abamectin for the management of the Asian citrus psyllid, *Diaphorina citri* (Hemiptera: Psyllidae). *Fla. Entomol.* **2008**, *91*, 87–100. [CrossRef]
7. Witten, I.H.; Frank, E.; Hall, M.A.; Pal, C.J. *Data Mining: Practical Machine Learning Tools and Techniques*; Morgan Kaufmann: Burlington, MA, USA, 2016.
8. Singerman, A.; Rogers, M.E. The economic challenges of dealing with citrus greening: The case of Florida. *J. Integr. Pest Manag.* **2020**, *11*, 3. [CrossRef]
9. Chen, X.-D.; Kaur, N.; Horton, D.R.; Cooper, W.R.; Qureshi, J.A.; Stelinski, L.L. Crude Extracts and Alkaloids Derived from Ipomoea-Periglandula Symbiotic Association Cause Mortality of Asian Citrus Psyllid *Diaphorina citri* Kuwayama (Hemiptera: Psyllidae). *Insects* **2021**, *12*, 929. [CrossRef]
10. Tian, F.; Rizvi, S.A.H.; Liu, J.; Zeng, X. Differences in susceptibility to insecticides among color morphs of the Asian citrus psyllid. *Pestic. Biochem. Physiol.* **2020**, *163*, 193–199. [CrossRef]
11. Lee, X.; Wong, C.; Coats, J.; Paskewitz, S.M. Semi-field evaluations of three botanically derived repellents against the blacklegged tick, *Ixodes scapularis* (Acari: Ixodidae). *bioRxiv* **2022**, *12*, 476114.
12. Kamaraj, C.; Gandhi, P.R.; Elango, G.; Karthi, S.; Chung, I.-M.; Rajakumar, G. Novel and environmental friendly approach; Impact of Neem (*Azadirachta indica*) gum nano formulation (NGNF) on *Helicoverpa armigera* (Hub.) and *Spodoptera litura* (Fab.). *Int. J. Biol. Macromol.* **2018**, *107*, 59–69. [CrossRef]
13. Achimón, F.; Areco, V.A.; Brito, V.D.; Peschiutta, M.L.; Merlo, C.; Pizzolitto, R.P.; Zygadlo, J.A.; Zunino, M.P.; Omarini, A.B. Plants as Bioreactors for the Production of Biopesticides. *Plants Bioreact. Ind. Mol.* **2023**, 337–366.
14. Chen, J. Biopesticides: Global Markets to 2022. *Rep. Code CHM029G* **2018**.
15. Tripathi, A.K.; Upadhyay, S.; Bhuiyan, M.; Bhattacharya, P. A review on prospects of essential oils as biopesticide in insect-pest management. *J. Pharmacogn. Phytother.* **2009**, *1*, 052–063.
16. Isman, M.B. Bridging the gap: Moving botanical insecticides from the laboratory to the farm. *Ind. Crops Prod.* **2017**, *110*, 10–14. [CrossRef]
17. Titouhi, F.; Amri, M.; Messaoud, C.; Haouel, S.; Youssfi, S.; Cherif, A.; Jemâa, J.M.B. Protective effects of three *Artemisia* essential oils against *Callosobruchus maculatus* and *Bruchus rufimanus* (Coleoptera: Chrysomelidae) and the extended side-effects on their natural enemies. *J. Stored Prod. Res.* **2017**, *72*, 11–20. [CrossRef]
18. Pradeepa, V.; Senthil-Nathan, S.; Sathish-Narayanan, S.; Selin-Rani, S.; Vasantha-Srinivasan, P.; Thanigaivel, A.; Ponsankar, A.; Edwin, E.S.; Sakthi-Bagavathy, M.; Kalaivani, K.; et al. Potential mode of action of a novel plumbagin as a mosquito repellent against the malarial vector *Anopheles stephensi*, (Culicidae: Diptera). *Pestic. Biochem. Physiol.* **2016**, *134*, 84–93. [CrossRef]
19. Rusin, M.; Gospodarek, J.; BINIAŚ, B. Effect of water extracts from *Artemisia absinthium* L. on feeding of selected pests and their response to the odor of this plant. *J. Cent. Eur. Agric.* **2016**, *17*, 188–206. [CrossRef]
20. Aslan, I.; Kordali, S.; Calmasur, O. Toxicity of the vapours of *Artemisia absinthium* essential oils to *Tetranychus urticae* Koch and *Bemisia tabasi*(Genn.). *Fresenius Environ. Bull.* **2005**, *14*, 413–417.

21. Husain, M.A.; Nath, D. *The Citrus Psylla: (Diaphorina citri, Kuw.) Psyllidae: Homoptera*; Government of India Central Pubilication Branch: Delhi, India, 1927.
22. Wenninger, E.J.; Stelinski, L.L.; Hall, D.G. Behavioral evidence for a female-produced sex attractant in *Diaphorina citri*. *Entomol. Exp. Appl.* **2008**, *128*, 450–459. [CrossRef]
23. Rizvi, S.A.H.; Tao, L.; Zeng, X. Chemical composition of essential oil obtained from *Artemisia absinthium* L. Grown under the climatic condition of Skardu Baltistan of Pakistan. *Pak. J. Bot* **2018**, *50*, 599–604.
24. Silva, J.A.; Hall, D.G.; Gottwald, T.R.; Andrade, M.S.; Maldonado, W.; Alessandro, R.T.; Lapointe, S.L.; Andrade, E.C.; Machado, M.A. Repellency of selected *Psidium guajava* cultivars to the Asian citrus psyllid, *Diaphorina citri*. *Crop Prot.* **2016**, *84*, 14–20. [CrossRef]
25. Barman, J.C.; Campbell, S.A.; Zeng, X. Exposure to Guava affects citrus olfactory cues and attractiveness to *Diaphorina citri* (Hemiptera: Psyllidae). *Environ. Entomol.* **2016**, *45*, 694–699. [CrossRef] [PubMed]
26. Hu, W.; Zheng, R.; Feng, X.; Kuang, F.; Chun, J.; Xu, H.; Chen, T.; Lu, J.; Li, W.; Zhang, N. Emergence inhibition, repellent activity and antifeedant responds of mineral oils against Asian citrus psyllid, *Diaphorina citri* (Hemiptera: Liviidae). *Int. J. Pest Manag.* **2023**, *69*, 27–34. [CrossRef]
27. Tiwari, S.; Stelinski, L.L.; Rogers, M.E. Biochemical basis of organophosphate and carbamate resistance in Asian citrus psyllid. *J. Econ. Entomol.* **2012**, *105*, 540–548. [CrossRef]
28. Walthall, W.K.; Stark, J.D. Comparison of two population-level ecotoxicological endpoints: The intrinsic (rm) and instantaneous (ri) rates of increase. *Environ. Toxicol. Chem. Int. J.* **1997**, *16*, 1068–1073.
29. Bowles, S.; Gintis, H. The evolution of strong reciprocity: Cooperation in heterogeneous populations. *Theor. Popul. Biol.* **2004**, *65*, 17–28. [CrossRef]
30. Anwar, M.I.; Sadiq, N.; Aljedani, D.M.; Iqbal, N.; Saeed, S.; Khan, H.A.A.; Naeem-Ullah, U.; Aslam, H.M.F.; Ghramh, H.A.; Khan, K.A. Toxicity of different insecticides against the dwarf honey bee, *Apis florea* Fabricius (Hymenoptera: Apidae). *J. King Saud Univ. Sci.* **2022**, *34*, 101712. [CrossRef]
31. Wang, Y.; Zhu, Y.C.; Li, W. Interaction patterns and combined toxic effects of acetamiprid in combination with seven pesticides on honey bee (*Apis mellifera* L.). *Ecotoxicol. Environ. Saf.* **2020**, *190*, 110100. [CrossRef]
32. Jaleel, W.; Yin, J.; Wang, D.; He, Y.; Lu, L.; Shi, H. Using two-sex life tables to determine fitness parameters of four *Bactrocera* species (Diptera: Tephritidae) reared on a semi-artificial diet. *Bull. Entomol. Res.* **2018**, *108*, 707–714. [CrossRef]
33. Rizvi, S.A.H.; Ling, S.; Tian, F.; Xie, F.; Zeng, X. Toxicity and enzyme inhibition activities of the essential oil and dominant constituents derived from *Artemisia absinthium* L. against adult Asian citrus psyllid *Diaphorina citri* Kuwayama (Hemiptera: Psyllidae). *Ind. Crops Prod.* **2018**, *121*, 468–475. [CrossRef]
34. Suladze, T.; Kintsurashvili, L.; Mshvildadze, V.; Todua, N.; Chincharadze, D.; Legault, J.; Vachnadze, N. Study of the Cytotoxic Activity of Alkaloid-Containing Fractions Isolated from Certain Plant Species Growing and Introduced in Georgia. *Exp. Clin. Med. Ga.* **2023**. [CrossRef]
35. da Silva Sa, G.C.; Bezerra, P.V.V.; da Silva, M.F.A.; da Silva, L.B.; Barra, P.B.; de Fátima Freire de Melo Ximenes, M.; Uchoa, A.F. Arbovirus vectors insects: Are botanical insecticides an alternative for its management? *J. Pest Sci.* **2023**, *96*, 1–20. [CrossRef]
36. Yang, Y.; Aghbashlo, M.; Gupta, V.K.; Amiri, H.; Pan, J.; Tabatabaei, M.; Rajaei, A. Chitosan nanocarriers containing essential oils as a green strategy to improve the functional properties of chitosan: A review. *Int. J. Biol. Macromol.* **2023**, *23*, 123954. [CrossRef]
37. Ibrahim, M.A.; Kainulainen, P.; Aflatuni, A. Insecticidal, repellent, antimicrobial activity and phytotoxicity of essential oils: With special reference to limonene and its suitability for control of insect pests. *Agric. Food Sci.* **2008**, *10*, 243–259. [CrossRef]
38. Park, J.-H.; Jeon, Y.-J.; Lee, C.-H.; Chung, N.; Lee, H.-S. Insecticidal toxicities of carvacrol and thymol derived from *Thymus vulgaris* Lin. against *Pochazia shantungensis* Chou & Lu., newly recorded pest. *Sci. Rep.* **2017**, *7*, 40902.
39. Karabörklü, S.; Ayvaz, A. A comprehensive review of effective essential oil components in stored-product pest management. *J. Plant Dis. Prot.* **2023**, *13*, 449–481. [CrossRef]
40. Ben Abada, M.; Soltani, A.; Tahri, M.; Haoual Hamdi, S.; Boushih, E.; Fourmentin, S.; Greige-Gerges, H.; Mediouni Ben Jemâa, J. Encapsulation of *Rosmarinus officinalis* essential oil and of its main components in cyclodextrin: Application to the control of the date moth *Ectomyelois ceratoniae* (Pyralidae). *Pest Manag. Sci.* **2023**, *79*, 2433–2442. [CrossRef]
41. Xie, F.; Rizvi, S.A.H.; Zeng, X. Fumigant toxicity and biochemical properties of (α+ β) thujone and 1, 8-cineole derived from *Seriphidium brevifolium* volatile oil against the red imported fire ant *Solenopsis invicta* (Hymenoptera: Formicidae). *Rev. Bras. Farmacogn.* **2020**, *29*, 720–727. [CrossRef]
42. Karalija, E.; Šamec, D.; Dahija, S.; Ibragić, S. Plants strike back: Plant volatiles and their role in indirect defence against aphids. *Physiol. Plant.* **2023**, *175*, e13850. [CrossRef]
43. Silva, M.S.; Patt, J.M.; de Jesus Barbosa, C.; Fancelli, M.; Mesquita, P.R.R.; de Medeiros Rodrigues, F.; Schnadelbach, A.S. Asian citrus psyllid, *Diaphorina citri* (Hemiptera: Liviidae) responses to plant-associated volatile organic compounds: A mini-review. *Crop Prot.* **2023**, *169*, 106242. [CrossRef]
44. Ling, S.; Rizvi, S.A.H.; Xiong, T.; Liu, J.; Gu, Y.; Wang, S.; Zeng, X. Volatile signals from guava plants prime defense signaling and increase jasmonate-dependent herbivore resistance in neighboring citrus plants. *Front. Plant Sci.* **2022**, *13*, 833562. [CrossRef] [PubMed]
45. Gallinger, J.; Rid-Moneta, M.; Becker, C.; Reineke, A.; Gross, J. Altered volatile emission of pear trees under elevated atmospheric CO_2 levels has no relevance to pear psyllid host choice. *Environ. Sci. Pollut. Res.* **2023**, *30*, 43740–43751. [CrossRef] [PubMed]

46. Ruan, C.-Q.; Hall, D.G.; Liu, B.; Duan, Y.-P.; Li, T.; Hu, H.-Q.; Fan, G.-C. Host-choice behavior of *Diaphorina citri* Kuwayama (Hemiptera: Psyllidae) under laboratory conditions. *J. Insect Behav.* **2015**, *28*, 138–146. [CrossRef]
47. Dong, Z.; Liu, X.; Srivastava, A.K.; Tan, Q.; Low, W.; Yan, X.; Wu, S.; Sun, X.; Hu, C. Boron deficiency mediates plant-insect (*Diaphorima citri*) interaction by disturbing leaf volatile organic compounds and cell wall functions. *Tree Physiol.* **2023**, *43*, 597–610. [CrossRef]
48. Hall, D.; Gottwald, T.; Nguyen, N.; Ichinose, K.; Le, Q.; Beattie, G.; Stover, E. Refereed manuscript. In Proceedings of the Florida State Horticultural Society; pp. 104–109.
49. Mann, R.; Rouseff, R.; Smoot, J.; Castle, W.; Stelinski, L. Sulfur volatiles from *Allium* spp. affect Asian citrus psyllid, *Diaphorina citri* Kuwayama (Hemiptera: Psyllidae), response to citrus volatiles. *Bull. Entomol. Res.* **2011**, *101*, 89–97. [CrossRef]
50. Mann, R.S.; Rouseff, R.L.; Smoot, J.; Rao, N.; Meyer, W.L.; Lapointe, S.L.; Robbins, P.S.; Cha, D.; Linn, C.E.; Webster, F.X. Chemical and behavioral analysis of the cuticular hydrocarbons from Asian citrus psyllid, *Diaphorina citri*. *Insect Sci.* **2013**, *20*, 367–378. [CrossRef]
51. Zheng, Y.; Zhang, J.; Li, Y.; Liu, Y.; Liang, J.; Wang, C.; Fang, F.; Deng, X.; Zheng, Z. Pathogenicity and Transcriptomic Analyses of Two "*Candidatus Liberibacter* asiaticus" Strains Harboring Different Types of Phages. *Microbiol. Spectr.* **2023**, *11*, e00754-23. [CrossRef]
52. Hosseinzadeh, S.; Heck, M. Variations on a theme: Factors regulating interaction between *Diaphorina citri* and "*Candidatus Liberibacter* asiaticus" vector and pathogen of citrus huanglongbing. *Curr. Opin. Insect Sci.* **2023**, *56*, 101025. [CrossRef]
53. Tiwari, S.; Stelinski, L.L. Effects of cyantraniliprole, a novel anthranilic diamide insecticide, against Asian citrus psyllid under laboratory and field conditions. *Pest Manag. Sci.* **2013**, *69*, 1066–1072. [CrossRef]
54. Assadpour, E.; Can Karaça, A.; Fasamanesh, M.; Mahdavi, S.A.; Shariat-Alavi, M.; Feng, J.; Kharazmi, M.S.; Rehman, A.; Jafari, S.M. Application of essential oils as natural biopesticides; recent advances. *Crit. Rev. Food Sci. Nutr.* **2023**, 1–21. [CrossRef]
55. Wuryantini, S.; Yudistira, R. The toxicity of the extract of tobacco leaf *Nicotiana tabacum* L, marigold leaf *Tithonia diversifolia* (HAMSLEY) and *Citrus japansche* citroen peel *Citrus limonia* against citrus psyllid (*Diaphorina citri* Kuwayama), the vector of citrus HLB disease. *IOP Conf. Ser. Earth Environ. Sci.* **2020**, *457*, 012039. [CrossRef]
56. Hall, D.G.; Borovsky, D.; Chauhan, K.R.; Shatters, R.G. An evaluation of mosquito repellents and essential plant oils as deterrents of Asian citrus psyllid. *Crop Prot.* **2018**, *108*, 87–94. [CrossRef]
57. Tian, F.; Li, C.; Wang, Z.; Liu, J.; Zeng, X. Identification of detoxification genes in imidacloprid-resistant Asian citrus psyllid (Hemiptera: Lividae) and their expression patterns under stress of eight insecticides. *Pest Manag Sci* **2019**, *75*, 1400–1410. [CrossRef]
58. Lee, S.; Peterson, C.J.; Coats, J. Fumigation toxicity of monoterpenoids to several stored product insects. *J. Stored Prod. Res.* **2003**, *39*, 77–85. [CrossRef]
59. Tak, J.-H.; Isman, M.B. Enhanced cuticular penetration as the mechanism for synergy of insecticidal constituents of rosemary essential oil in *Trichoplusia ni*. *Sci. Rep.* **2015**, *5*, 12690. [CrossRef]
60. Pavela, R. Limitation of plant biopesticides. In *Advances in Plant Biopesticides*; Springer: Berlin/Heidelberg, Germany, 2014; pp. 347–359.
61. Bandi, S.M.; Mishra, P.; Venkatesha, K.; Aidbhavi, R.; Singh, B. Insecticidal, residual and sub-lethal effects of some plant essential oils on *Callosobruchus analis* (F.) infesting stored legumes. *Int. J. Trop. Insect Sci.* **2023**, *43*, 383–395. [CrossRef]
62. Ciobanu, A.; Landy, D.; Fourmentin, S. Complexation efficiency of cyclodextrins for volatile flavor compounds. *Food Res. Int.* **2013**, *53*, 110–114. [CrossRef]
63. Costa, J.A.V.; Freitas, B.C.B.; Cruz, C.G.; Silveira, J.; Morais, M.G. Potential of microalgae as biopesticides to contribute to sustainable agriculture and environmental development. *J. Environ. Sci. Health Part B* **2019**, *54*, 366–375. [CrossRef]

Disclaimer/Publisher's Note: The statements, opinions and data contained in all publications are solely those of the individual author(s) and contributor(s) and not of MDPI and/or the editor(s). MDPI and/or the editor(s) disclaim responsibility for any injury to people or property resulting from any ideas, methods, instructions or products referred to in the content.

Article

A Comparison of IPM and Organic Farming Systems Based on the Efficiency of Oophagous Predation on the Olive Moth (*Prays oleae* Bernard) in Olive Groves of Southern Iberia

José Alfonso Gómez-Guzmán [1], José M. Herrera [2,3], Vanesa Rivera [2], Sílvia Barreiro [2], José Muñoz-Rojas [2,4], Roberto García-Ruiz [1] and Ramón González-Ruiz [1,*]

[1] Department of Animal Biology, Plant Biology and Ecology, University Institute of Research on Olive Groves and Olive Oils, Universidad de Jaén, 23071 Jaén, Spain
[2] Mediterranean Institute for Agriculture, Environment and Development, University of Évora, 7000-651 Évora, Portugal
[3] Department of Biology-IVAGRO, Universidad de Cádiz, 11510 Puerto Real, Spain
[4] Department of Geosciences, University of Évora, 7006-554 Évora, Portugal
* Correspondence: ramonglz@ujaen.es; Tel.: +34-9-5321-2499

Abstract: The olive moth, *Prays oleae* (Bernard, 1788) (Lep., Praydidae), is one of the most common insect pests affecting the olives groves of the Mediterranean basin. Current farming practices are largely oriented to optimize the effectiveness of beneficial insects, among which the common green lacewings (Neur., Chrysopidae) stand out. Two different types of management models, organic and IPM, were compared in this study, which was conducted in olive groves in the regions of Andalucía (Spain) and Alentejo (Portugal). During 2020 and 2021, fruit samples were periodically collected, analyzing the population parameters (POP) and potential attack on the fruit (%PA), as well as the predatory impact (%PRED), which has allowed the estimation of the final attack (%FA), and derived fruit recovery rates (%REC). The results show that in organic olive groves of both countries, the infestation parameters (POP, %PA) were significantly higher than in IPM ones. However, predation rates were also higher in organic olive groves, which resulted in REC rates of between 75% and 80%, reducing FA rates to values of approximately 10% and 20% in Portugal and Spain, respectively. In contrast, in the IPM olive groves, significantly lower predation values were recorded, with lower REC rates than in the organic olive groves; the rates were very similar in both countries (ca. 54%), which led to a higher percentage of fruit loss (%FA) equivalent to 22% (Portugal) and to 34% (Spain). This paper discusses potential drivers influencing differences in the population values and percentages of infestation by *P. oleae* observed, as well as the differences in the final attack rates between olive groves of both countries, subject to the same type of agricultural management.

Keywords: olive farming; Chrysopidae; *Chrysoperla carnea* complex; olive losses; oophagous predation

1. Introduction

The olive moth, *Prays oleae* (Bernard, 1788) (Lep., Praydidae), is one of the most common insects that damages olives in the producing countries of the Mediterranean basin [1–3] in the south of the Iberian Peninsula, where approximately 50% of the world's olive oil is produced [4]. Economic losses caused by this pest can exceed 40% of the harvest [2,5,6]. These are mainly caused by the carpophagous generation of the olive moth [7] whose larvae feed on the tissue inside the olive stone during the summer. Once developed, the larva emerges at the end of summer from the fruit through a hole in the apical zone, causing the fall of the affected fruits [2,5,6].

Regarding the interaction between *P. oleae* and its natural enemies, Chrysopids (Neuroptera) are among its most voracious predators [8–14]. These predators have long attracted the attention of applied entomologists, for they are good candidates for use in integrated pest management (IPM) programs [8]. This is due to their wide geographic distribution, their wide spectrum of host plants and prey [15], their relatively easy mass production [16], the possibility of protecting lacewing numbers by overwintering chambers [17], and their ability to develop pesticide-tolerant populations [18]. Given these advantages, members of the Association of Applied Insect Ecologists (AAIE) placed *Chrysoperla* ssp. as the most important lacewings, unrivaled among all other commercially obtainable predatory species [19].

The taxonomic status of the common green lacewing, which was formerly named as *Chrysoperla carnea* (Stephens, 1836) (Neur., Chrysopidae), has undergone major changes, and instead of a polymorphic single species, it is accepted as a complex of sibling cryptic species, the *Chrysoperla carnea* complex, or *carnea* group [20,21]. The existence of the following cryptic species has been assumed: (1) *Chrysoperla affinis* (Stephens, 1836) former *Ch. kolthoffi* (Navás, 1927) [20], (2) *Chrysoperla lucasina* (Lacroix, 1912) [21], (3) *Chrysoperla carnea* sensu stricto [20] and *Chrysoperla pallida* Henry et al., 2002 [22], and (4) *Chrysoperla agilis* Henry et al., 2003 [23]. In southern Spain, the presence of these cryptic species has been reported in Andalusian olive groves, of which the later species, *Ch. agilis,* was dominant (>90%) [24]. In olive groves, *carnea* complexes are polyphagous and effective predators in the natural control of *P. oleae* [1,24].

Several methods are used to control moth populations, although most of them are based on the application of synthetic insecticides, of which the use is not as difficult as it could be, especially because these chemicals are highly harmful to natural enemies [6]. The adverse effects are especially notable in crops under conventional farming practices, in which the frequency of use of synthetic insecticides is much higher than any other type of agricultural management. This farming system is followed by the largest proportion of Spanish olive groves (70%) [25], as is also the case in Portugal. In addition, in these crops, plant biodiversity is extraordinarily low, as the herbaceous plant cover is often eradicated by the regular application of herbicides.

In view of the adverse effects and environmental risks of conventional management, integrated pest management (IPM) allows notable environmental improvements [26]. This farming system is followed in approximately 25% of Spanish olive groves [25] where the objective is to reconcile chemical and natural control elements through a reduction in insecticide application and an attempt to adapt the crop in an agroecosystem more favorable for beneficial insects. Among the innovative elements incorporated in the IPM olive groves is the promotion of a herbaceous vegetable cover that protects the soil from erosion and contributes to maintaining the populations of beneficial insects. In addition, IPM drives the establishment of population thresholds, which allows farmers to discern the need to apply synthetic pesticides, which leads to considerable savings in the use of synthetic pesticides. However, in practice, important aspects of IPM practices remain largely unresolved, such as the integration of the population levels of natural enemies/antagonists in decision-making, as well as improving pest management programs through the implementation of less disruptive tactics. The integration of these aspects must consider both the damages caused in the control of the target pest (called vertical or first level), as well as the indirect effects on the remaining pests (considered as a second or horizontal level) [27]. However, regarding IPM crops, environmentally safer farming systems have arisen, such as organic olive growing (syn. ecological) and biodynamic olive growing, which can be considered the greatest exponents and best approaches towards sustainability [28]. These two types of farming systems, although with different origins, are largely similar, representing a holistic view of nature based on a cyclic understanding of crop production, along with a return to traditional agricultural principles [29]; the foundations are based on a mixture of ancient agronomic philosophies, empirical observations and scientific approaches [28]. Among the main differences between them, biodynamic crops incorporate specific fermented herb preparations as compost additives. In both systems, current standards for certified production have a legally regulated procedure, outlining the use of soil building activities and natural pest management [30] through a plan that allows the application of pesticides of natural origin. The complete suppression of synthetic pesticides is reflected in organic crops through a greater diversity and abundance of natural enemies [31,32], and frequently results in an acceptable level of pest control [33]. However, as demonstrated by Scherber et al. [34], increased biodiversity in organic farms does not always suffice to adequately control pests, as it enforces the farmer to use organically accepted pesticides, which negatively affect biodiversity. Among the drawbacks, it has been highlighted that the methods used in organic farming can be more expensive than conventional chemical-based farming practices, although these estimates often do not take into account the high environmental costs associated with the use of conventional pesticides [35]. As a result, the organic industry is not yet at a level that allows it to compete with synthetic pesticides, so the proportion of organic olive farming represents only a minority (4% organic sensu stricto and 1% organic–biodynamic in the Iberian Peninsula) [25].

Organic farming systems thus represents an approach more aligned with natural conditions than IPM, which a priori should lead to a higher level of biodiversity and sustainability standards. As mentioned before, although increased biodiversity does not always suffice in adequately controlling pests, an increase in predation rates of endemic pests such as the olive moth would be expected. The objective of this study is to make an accurate estimate of the predatory impact of the common green lacewings on the carpophagous generation of *P. oleae*. This may allow for an accurate comparison of the impact of the pest on the harvest of the two farming systems in olive groves of Spain and Portugal, and provide practical elements for their better characterization in terms of sustainability.

2. Materials and Methods

2.1. Description of the Study Area

The study was carried out in 10 olive groves in southern Portugal (Alentejo region, Évora and Beja districts) and in 12 olive groves in southern Spain (Andalusia region, Jaén and Granada provinces) during the oviposition period of the anthophagous generation of *P. oleae* (June/July 2021). The start of the oviposition period took place about 7 days after the start of the fruiting process (phenological stage G) [36] that occurred in both countries in the last week of May. The selected olive groves (Figure 1) were subject to the two different types of management practices analyzed: IPM and organic management (ORG). The main characteristics of the crop and the control measures applied are summarized in Tables 1 and 2. It should be noted that, from Portugal, two olive groves with biodynamic olive growing were included in the study. These are the ones indicated (*) in Table 2. Since this type of olive management practice is very similar to organic management, these 2 olive groves were included within the cluster of organic olive groves.

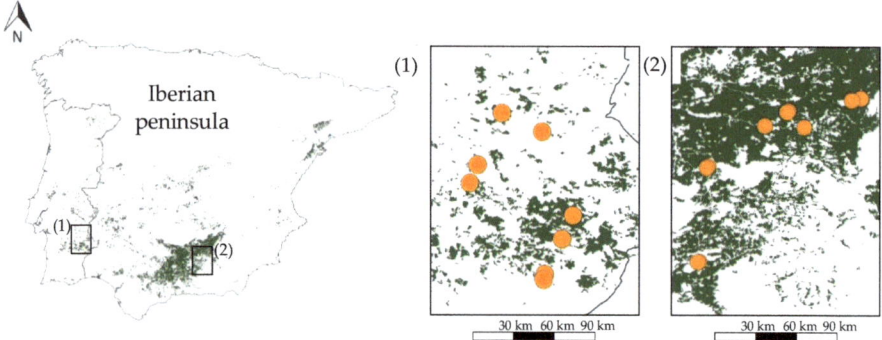

Figure 1. Location of the olive groves selected in this study. The green spots on the map correspond to the areas of land in Portugal (1) and Spain (2) used for olive cultivation. Orange circles indicate the location of the olive groves selected for the study.

Table 1. Geographical location of the olive groves (IPM and organic management) considered in the south of Spain. The main characteristics of each of them (variety, area, age of the trees, plantation density) and the chemical control measures applied to control herbaceous plant cover, as well as against the carpophagous generation of *P. oleae*, are indicated.

				Variety	Area (ha)	Age (Years)	Density (Trees × ha)	Herbicide Active Principle (Dose); Date	Insecticide (Olive Moth) Anth. Generation
SPAIN	IPM	Baeza	37°55′29.9″ N 3°24′10.9″ W	"Picual"	47.9	150	100	Glyphosate 36% (1.5 L/ha) + Flazasulfuron 25% (0.06 kg/ha); March & October	Deltametrin 10% (0.125 L/ha); April
		Deifontes	37°23′11.8″ N 3°38′09.9″ W	"Picual"	15.8	30	150		
		Pegalajar	37°45′03.2″ N 3°37′31.2″ W	"Picual"	24.3	80	80		
		Úbeda	37°58′57.7″ N 3°19′04.1″ W	"Picual"	9.4	40	100		
		Villacarrillo I	38°03′05.1″ N 3°01′28.3″ W	"Picual"	7.1	200	90		
		Villacarrillo II	38°02′31.0″ N 3°03′46.0″ W	"Picual"	4.8	100	100		
	ORGANIC	Baeza	37°55′40.0″ N 3°14′47.9″ W	"Picual"	23.6	100	100	–	–
		Deifontes	37°23′04.8″ N 3°38′30.2″ W	"Picual"	72.2	25	150		
		Pegalajar	37°45′21.8″ N 3°37′08.2″ W	"Picual"	15.6	40	120		
		Úbeda	37°59′12.2″ N 3°18′43.9″ W	"Picual"	58.8	40	100		
		Villacarrillo I	38°03′03.3″ N 3°01′33.0″ W	"Picual"	5.3	200	90		
		Villacarrillo II	38°02′29.5″ N 3°03′43.6″ W	"Picual"	1.8	100	100		

Table 2. Geographical location of the olive groves (IPM and organic management) considered in the south of Portugal. The main characteristics of each of them (variety, area, age of the trees, plantation density), and the chemical control measures applied to control the herbaceous plant cover, as well as against the anthophagous/carpophagous generations of *P. oleae*, are indicated. An asterisk (*) indicates organic–biodynamic crops.

				Variety	Area (ha)	Age (Years)	Density (Trees × ha)	Herbicide Active Principle (Dose); Date	Insecticide (Olive Moth) Anth. Generation	Carp. Generation
PORTUGAL	IPM	Évora	38°26′56.0″ N 7°41′22.2″ W	"Arbequina"	250	10	1667	Glyphosate 36% (3 L/ha); March and October	Lambda cihalotrin 15% (1.3 L/ha); April	Lambda cihalotrin 15% (1.3 L/ha); May–June
		Pias	38°01′30.8″ N 7°29′43.9″ W	"Cobrançosa"	140.5	20	205			
		Vidigueira I	38°10′01.5″ N 7°44′23.5″ W	"Cobrançosa"	130	16	285			
		Vidigueira II	38°11′19.8″ N 7°49′13.7″ W	"Cobrançosa"	163	29	200			
		Vidigueira III	38°11′01.1″ N 7°47′45.2″ W	"Arbequina"	163	9	600			
	ORGANIC	Portel I	38°16′26.9″ N 7°46′24.6″ W	"Galega"	35	100	139	–	–	Kaolin (35 kg/ha); May–June
		Portel II	38°18′26.9″ N 7°43′39.7″ W	"Galega" + "Blanqueta"	35	100	139			
		Reguengos de Monsaraz	38°23′10.7″ N 7°32′58.5″ W	"Cobrançosa"	100	13	285			
		Serpa I *	37°54′08.4″ N 7°32′24.2″ W	"Arbequina"	200	3	1770			
		Serpa II *	37°53′04.7″ N 7°32′38.9″ W	"Cobrançosa" + "Arbequina"	100	5	300			

2.2. Experimental Design

In Spain, 6 pairs of olive plots were chosen and within each pair one plot corresponds to organic management and the second to IPM. In this way, we intended to establish paired comparisons in each study area. In contrast, the distribution of management types in Alentejo (Portugal) did not allow for the establishment of paired plots. To cope with this, 5 olive groves were selected with organic farming and 5 with IPM farming, but these plots were not directly adjacent (Table 2). Consequently, in Portugal, the comparisons of the different parameters between the two types of agricultural management models were made considering the data corresponding to the set of crops of each type of farming.

2.2.1. Sampling of Fruit and Determination of the Attack Parameters of *P. oleae*

The study was carried out during the oviposition period of the anthophagous generation of *P. oleae*. The egg population was monitored by periodic fruit sampling. The first fruit sampling in the olive groves of both countries was carried out at the beginning of June, 7 days after the beginning of the oviposition period and about 15 days after the start of the fruit formation process (phenological stage G). From the first sampling date onwards, fruit samples were collected at 15-day intervals so that by the end of the oviposition period, a total of 4 samples had been collected in each olive grove. In Spain, on each sampling date, 6 olive trees were randomly selected from each olive grove, while in Portugal, from each olive grove, 3 olive trees were randomly selected. From each selected olive tree, 100 fruits were collected (25 fruits of each cardinal orientation). The fruits were and placed in opaque plastic vials, protecting them from solar radiation, and were taken to the laboratory where they were temporarily stored in a cold room (T < 4 °C).

The olives were observed using a stereomicroscope (LEICA, M205C, WETZLAR, GERMANY), taking note of the number of *P. oleae* eggs for each of them and differentiating between live eggs, predated eggs and hatched eggs according to the indications by Arambourg [8] and Ramos and Ramos [37]. Once all the fruits had been examined, the following parameters were calculated for each olive grove on each sampling date:

- Population Index (POP): Total number of eggs/100 fruits. This density value reflects the relative density of ovideponent females in the cultivated area.
- Potential Attack (%PA): Number of fruits with any type of eggs (live, predated or hatched) × 100/total number of fruits observed. It represents a value equivalent to the fruit drop that *P. oleae* would cause in case of the total absence of oophagous predation activity.
- Hatching rate: Number of hatched eggs × 100/(sum of live and hatched eggs).
- Predation rate (%PRED): Number of predated eggs/(sum of live eggs + predated + hatched). This allows for the assessment of predatory activity, an index of the activity of predatory eggs.
- Final Attack (%FA): Number of fruits that contain at least one hatched egg × 100/(number of fruits observed). Given that the loss of fruit caused by *P. oleae* is exclusively due to the emergence of the larvae inside the attacked fruits once their development is complete, only hatched eggs are considered for its calculation. Therefore, it may be considered that the latter have survived the predatory action of natural enemies. For the calculation of the %FA value, therefore, live eggs were not taken into account, since these are likely to hatch or be predated. The entomophagous action allows discarding a proportion of the population of *P. oleae* eggs; however, in the opposite case, its hatching involves the establishment of the larva in the endocarp of the fruit and his subsequent fall. %FA describes the magnitude of this drop, providing a realistic estimation of the impact of *P. oleae* on the harvest.
- Fruit Recovery (%REC): Number of fruits in which all the eggs have been predated × 100/(number of fruits that have contained eggs). This parameter indicates the real effectiveness of predation, since it corresponds to the percentage of fruits in which all the eggs have been predated. Once all *P. oleae* eggs have been eliminated, fruits are considered as recovered, so for practical purposes, the recovery percentage is a parameter that indicates the real effectiveness of oophagous predation by lacewings. It is important to note that this value does not always correspond to the %PRED, except in those cases in which the number of fruits attacked present infestation densities equal to 1 egg/fruit.

2.2.2. Sampling of Lacewings

During the summer of 2021, lacewings adults were sampled in the olive groves of southern Spain. In each pair of plots, 6 traps were placed (3 for IPM and 3 for organic) which were installed in central olive trees within each plot, at a height of 1.5 m, in the southern side of the olive trees. This was performed at a rate of one trap per tree and separated by a minimum distance of 50 m, using McPhail traps baited with a solution of diammonium phosphate (4% w/v). The traps were installed once the oviposition process of *P. oleae* had begun, renewing them at 10-day intervals. A sieve was used to collect the insects captured in the traps, which were later placed in glass vials containing an aqueous dilution of 70% ethanol. In the laboratory, a proportion of the captured individuals, as permitted by their conservation status, were determined taxonomically, recording the capture numbers for each species. According to previous studies, this type of trap is attractive for lacewings, although the largest proportion of captured individuals corresponds to female individuals (>90%) [12].

2.3. Statistical Analysis

For statistical analysis, the Statgraphics Centurion XVIII statistical package was used. The normality of the distributions was verified using the Shapiro–Wilk normality test in the case of a sample size of more than 50 data, while for sample sizes of less than 50 data, the Kolmogorov–Smirnov test (K–S, with the Liliefors correction) was applied since the power of the Shapiro–Wilk test is lower for a small sample size [38]. Levene's test was then applied to evaluate the homogeneity of variance. To determine the existence of significant differences between the parameters calculated in the IPM and organic olive groves (POP, %PA, %PRED, %FA and %REC), an analysis of variance (ANOVA) was performed. To determinate the level of statistical significance, Tukey's HSD (honestly significant difference) test was applied.

3. Results

3.1. Analysis of the Attack Parameters of P. oleae

3.1.1. Population Index (POP)

In the olive groves of southern Spain, the POP values showed statistically significant differences between the IPM and organic farming systems ($p < 0.01$ in five of the six pairs of plots), with greater records in the organic plots (Figure 2A,A'); at the end of the oviposition period values ranged between 106.7 and 248.0 egg/100 fruits (X = 176.1; SD = 46.8), while in the IPM plots, these values ranged between 84.3 and 181.7 egg/100 fruits (X = 142.9; SD = 35.1).

In the olive groves of southern Portugal, POP was generally lower than those described for southern Spain (Figure 3A,A'), ranging, at the end of the oviposition period, between 48.0 and 230.0 egg/100 fruits in organic olive groves (average of 121.2 egg/100 fruit; SD: 66.8 egg/100 fruit) and from 49.3 to 94.0 (average of 76.8; SD:19.9) in IPM olive groves. As was also the case in the Spanish olive groves, statistically significant differences were recorded between POP values for different olive farming systems, with greater values recorded in organic olives ($p < 0.001$; Table 3).

Table 3. Mean values and standard deviation rates of the Population Index (POP), Potential Attack (%PA), oophagous predation (%PRED), Final Attack (%FA) and Fruit Recovery (%REC) in the organic and IPM olive groves selected in Spain and Portugal (last sampling date). The level of statistical significance (ANOVA) is also indicated.

	SPAIN						PORTUGAL					
	IPM		ORG		ANOVA		IPM		ORG		ANOVA	
	Mean	(SD)	Mean	(SD)	F	*p*-Value	Mean	(SD)	Mean	(SD)	F	*p*-Value
POP	142.9	(35.1)	176.1	(46.8)	12.8	$p < 0.001$	76.8	(19.9)	121.2	(66.8)	19.6	$p < 0.05$
%PA	80.2	(12.3)	88.8	(7.3)	10.1	$p < 0.01$	58.3	(11.2)	71.0	(19.9)	12.1	$p < 0.05$
%PRED	68.1	(3.7)	85.8	(3.1)	80.6	$p < 0.001$	56.8	(4.9)	86.5	(4.1)	107.8	$p < 0.001$
%FA	34.0	(5.5)	19.7	(2.7)	32.5	$p < 0.001$	22.0	(8.0)	9.3	(4.2)	9.9	$p < 0.05$
%REC	53.8	(3.2)	76.2	(2.2)	192.6	$p < 0.001$	53.8	(5.5)	80.2	(7.2)	42.8	$p < 0.001$

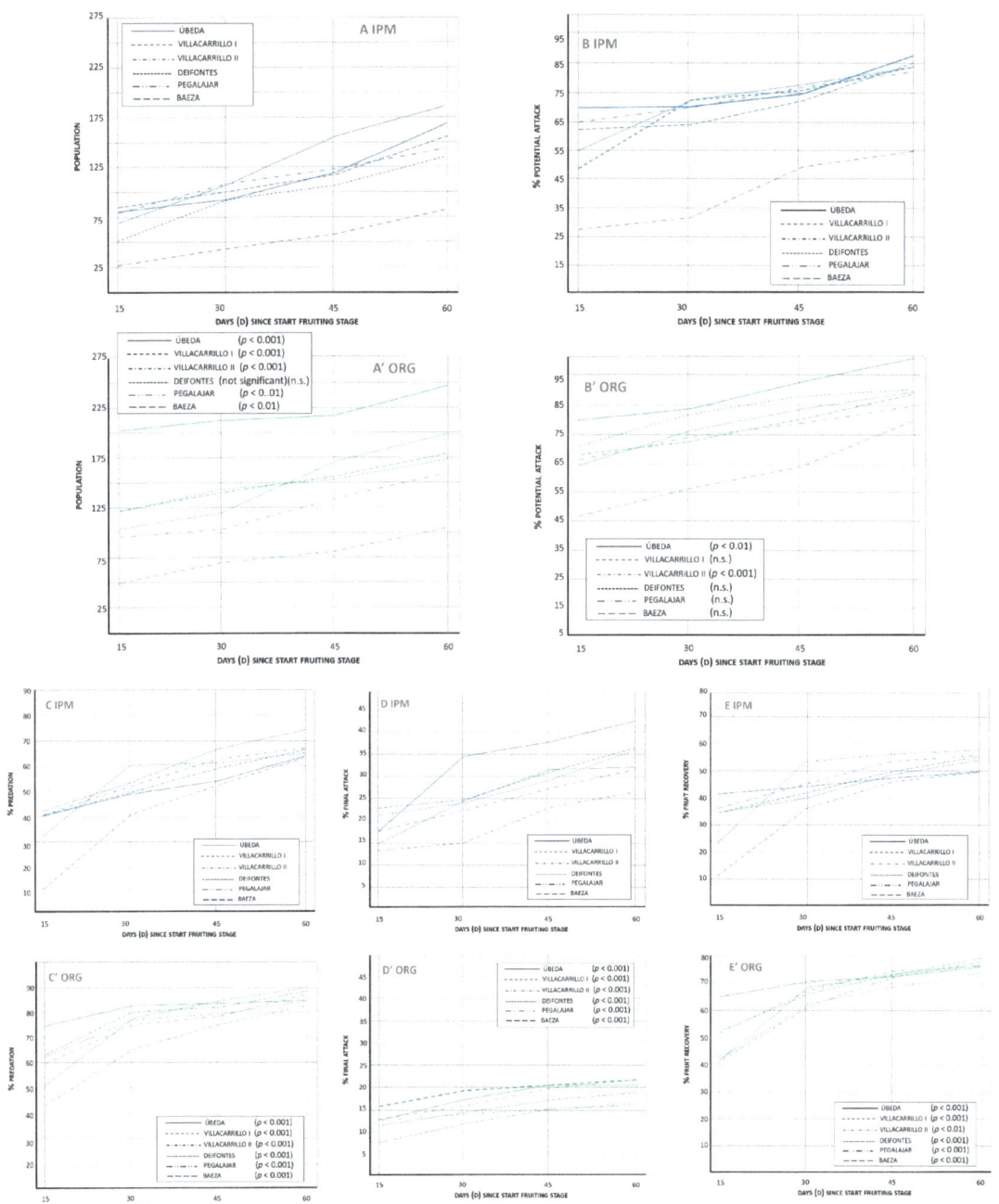

Figure 2. *P. oleae* attack parameters in the paired plots in southern Spain, corresponding to the two agronomic managements models: IPM (**A**–**E**) and organic (**A'**–**E'**). Population index (POP, Graphs **A**,**A'**). Potential Attack (%PA, Graphs **B**,**B'**); Predation (%PRED, Graphs **C**,**C'**); Final Attack (%FA, Graphs **D**,**D'**); % Recovery (%REC, Graphs **E**,**E'**). The graphs corresponding to the organic olive groves show the results of the analysis of variance (ANOVA) conducted to determine the levels of statistical significance between olive farming systems at the last sampling date.

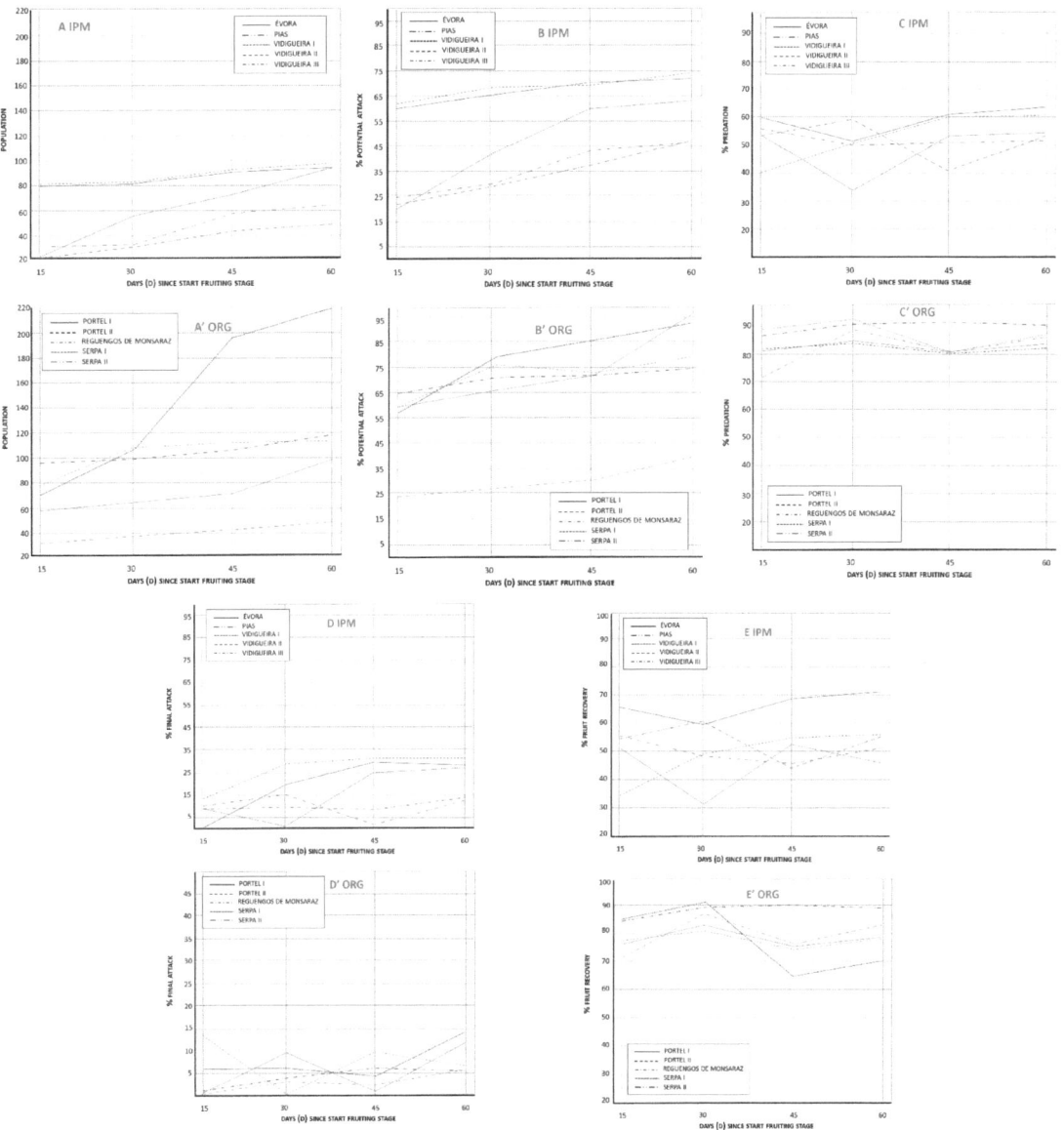

Figure 3. Parameters of *P. oleae* attack in olive groves under different agronomic management models: IPM (Graphs **A–D**) and organic (Graphs **A′–D′**) selected in southern Portugal. Population index (POP, Graphs **A,A′**). Potential Attack (%PA, Graphs **B,B′**); Predation (%PRED, Graphs **C,C′**); Final Attack (%FA, Graphs **D,D′**); % Recovery (%REC, Graphs **E,E′**). Among the organic olive groves, the olive groves corresponding to organic–biodynamic management are marked with red lines.

The sequential analysis of fruit sampling indicates that POP values increased gradually throughout the oviposition period of anthophagous generation, following a linear regression model (Figure 4A).

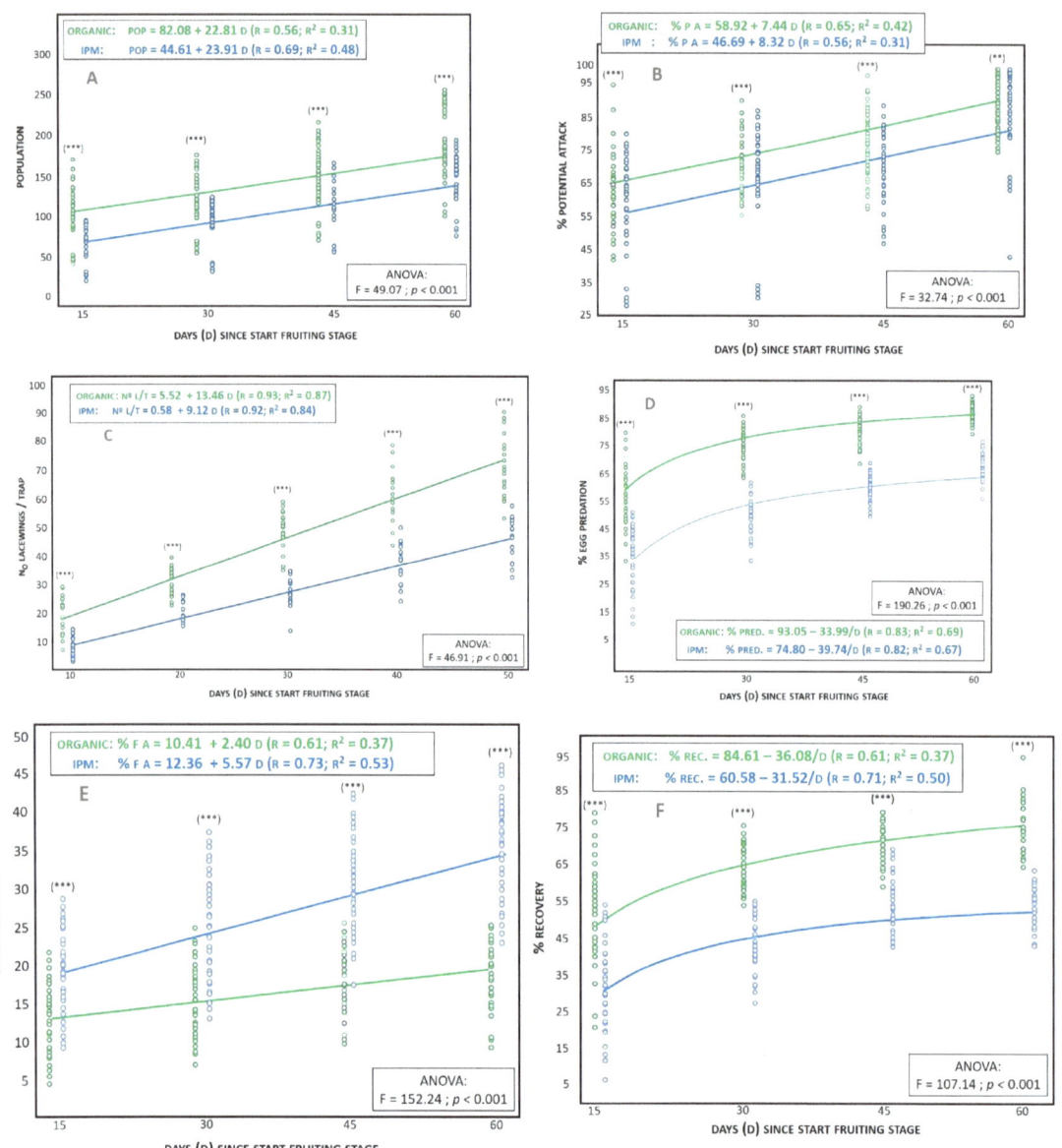

Figure 4. Regression curves of *P. oleae* attack parameters in pairs of plots with organic (green) and IPM (blue) management selected in southern Spain during the oviposition period. Each circle represents the value obtained in one of the olive trees selected from each plot on each sampling date. Population index (POP, (**A**)); Potential Attack (%PA, (**B**)). Number of captures of green lacewings in Mcphail traps (**C**). Egg predation (%PRED, (**D**)). Final Attack (%FA, (**E**)) and Fruit Recovery (%REC, (**F**)). With asterisks, the levels of statistical significance (ANOVA; ** $p < 0.01$; *** $p < 0.001$) between sampling series corresponding to organic and IPM plots, for each sampling date, are indicated.

Similar to what was obtained in the south of Spain, the %FA values of the olive groves of the south of Portugal presented statistically significant differences between IPM and organic systems (Figure 3D,D') (Table 3; $p < 0.001$), with the final %FA in the IPM ranging between 12.0% and 29.0% (X = 22.0%; SD = 8.0). In the Portuguese organic olive groves, %FA values ranged between 5.3% and 14.7%, with an average value of 9.3% (SD = 4.2).

3.1.2. Potential Attack (%PA)

Among the paired plots of Spain, %PA in the IPM plots ranged at the end of the oviposition period between 55.3% and 88.7% of the fruits (average of 80.2%; SD = 12.3%). These values were even longer in the organic plots—between 78.7% and 99.6% of the fruits—with a mean value of 88.8%, (SD = 7.3%). The analysis of the sequential fruit sampling showed that the %PA increased steadily and gradually throughout the oviposition period, which through regression analysis could be adjusted to a linear function (Figure 4B). In the organic plots, the %PA values were significantly higher in two out of the six pairs of plots (Figure 2B,B'). Equally, statistically significant differences also became evident when considering the final data (at the end of the oviposition period) from the six pairs of plots, considered as a whole (Table 3; $p < 0.01$).

%PA in olive groves in southern Portugal were approximately 20% lower than those recorded in southern Spain. In the IPM groves, the potential fall of the fruit caused by *P. oleae*—in the absence of oophagous predation—could have reached values between 46.3% and 70.3% (X = 58.3%; SD = 11.2%) at the last sampling date. Similar to what was reported for Spain, %PA in the Portuguese organic groves were significantly higher than in IPM orchards, ranging between 39.0% and 93.0% (X = 71.0%; SD = 19.9%) (Table 3; $p < 0.05$) at the last sampling date. As shown in Figure 3B,B', in the Portuguese olive groves with organic–biodynamic management, both the %PA values and their development during the oviposition period do not differ from the rest of the organic olive groves.

3.1.3. Lacewing Diversity and Egg Predation (%PRED)

Among the individuals captured in the McPhail traps, a sample of 25% of the captured lacewings were taxonomically determined, resulting in the identification of six species. Most (86%) corresponded to *Ch. agilis*, 4% to *Pseudomallada prasinus* (Burmeister, 1839), 3% to *Pseudomallada flavifrons* (Brauer, 1851), 3% to *Ch. affinis*, 2% to *Ch. lucasina* and 2% to *Chrysopa viridana* Schneider, 1845. Since *Ch. agilis* is the dominant species, its impact on olive moth eggs is the greatest amongst all lacewing species considered [24,39].

As shown in Figure 2C,C', the percentage of oophagous predation in Spanish olive groves increases rapidly throughout the oviposition period, reaching maximum values close to 90%, adjusting this variation to an inverse X function (Figure 4D). Predation values were significantly higher in plots under organic management (Figure 2C,C'), ranging at the end of the oviposition period between 82.1% and 89.6%, with a final average of 85.8% (SD: 3.1%). Statistically lower values were recorded in the IPM olive groves (Table 3; $p < 0.001$), which ranged between 64.7% and 75.1%, with an average of 68.1%, (SD = 3.7%).

Similarly to the olive groves of southern Spain, in southern Portugal there are also differences in the predation rates under both management models (Figure 3C,C'). These were significantly higher in organic farming (Table 3; $p < 0.001$), ranging between 82.3% and 92.9% (X = 86.5%; SD = 4.1%). As observed in %PA, %PRED in the groves with organic–biodynamic management in southern Portugal showed very similar trends to the rest of the organic ones (Figure 3C,C'). In contrast to the organic ones, %PRED in IPM Portuguese olive groves was, on average, 30% lower, ranging between 52.0% and 63.1%, resulting at the end of the oviposition period in a final average value of 56.8% (SD = 4.9) (Table 3).

3.1.4. Final Attack (%FA)

As indicated above, the attacked fruits will be lost at the end of the summer as a consequence of the emergence of the larva once its development is complete. In the IPM groves of southern Spain, %FA at the end of the oviposition period ranged between 26.0% and 42.3% of the fruits (X = 34.0%; SD = 5.5). In organic plots, %FA was significantly lower in every one of the pairs of plots (Figure 2D,D') ranging at the end of the oviposition period between 16.3% and 22.7% (X = 19.7%; SD = 2.7), which represents a decrease of 15% with respect to the IPM plots (Table 3; $p < 0.001$).

3.1.5. Fruit Recovery (%REC)

In the olive groves of southern Spain, the percentages of fruit in which eggs were present (with respect to the total number of infested fruits) had been entirely eliminated by the predatory action of the lacewings (%REC), ranging, at the end of the oviposition period, between 49.7% and 56.9% in the IPM plots, with an average value of 53.8% (SD = 3.2). Compared with the IPM plots, the %REC in the organic plots was significantly higher in each of the pairs of plots (Figure 2E,E'), ranging between 73.1% and 79.2%, with an average value of 76.2% (SD: 2.2) (Table 3). This represents an increase of approximately 22% with respect to the IPM groves.

As with the predation parameter, the percentage of fruit recovered in Spanish olive orchards increases rapidly throughout the oviposition period, with this variation fitting an inverse X function (Figure 4F).

Regarding the olive groves of southern Portugal with IPM management, %REC (Figure 2B,B') ranged, at the end of the oviposition period, between 46.1% and 61.1%, with an average value of 53.8% (SD = 5.5). In the plots under organic management, significantly higher data were recorded (Table 3; $p < 0.001$), ranging between 70.2% and 89.4%, with an average value of 80.2% (SD: 7.2%) (Table 3).

4. Discussion

From the results of the comparative study of the impact of organic/IPM farming systems on the entomophagous activity of lacewings (oophagous predators of *P. oleae*) during their carpophagous generation, a series of aspects ought to be considered. Among the parameters analyzed, it is significant to note that in olive groves of both countries there are differences between both types of management. With regard to POP and %PA, these represent a measure of the relative density of *P. oleae* populations, thus acting as indicators of the degree of infestation in the fruit, which is independent of environmental and ecological variability, since in both countries the highest values were recorded in olive groves cultivated with organic farming. Most likely, this difference is a consequence of the fact that the control procedures applied in organic olive groves against the carpophagous generation of *P. oleae* have either been completely suppressed (as is the case with organic olive groves in southern Spain in this study), or are based on environmentally friendly non-chemical control measures, such as the application of a thin layer of kaolin, which is increasingly being applied for the control of important olive grove pests [40,41]. This measure represents a physical barrier of kaolin that would limit the oviposition of females, as was detected in the organic olive groves in southern Portugal, where lower rates of POP and %PA have been recorded in relation to Spanish organic olive groves. In agreement with these observations, Pascual et al. [40] indicate a significant reduction in the infestation of important olive grove pests, such as the olive black scale *Saissetia oleae* (Olivier) (Hem., Coccidae), and the olive fruit fly *Bactrocera oleae* (Rossi, 1790) (Dip., Tephritidae), through the application of kaolin particle films.

Regarding the percentages of oophagous predation (%PRED) in both countries, they were detected as being much higher in organic olive groves (reaching rates of 85–87%) in relation to IPM olive groves, where they were on average 30% lower. This difference is attributable to the lighter ecological pressure of the control measures applied in organic olive groves. In this sense, since no control measures had been applied to the olive moth in organic olive groves in southern Spain, it follows that kaolin, in addition to significantly reducing the rate of *P. oleae* infestation in the fruits, appears to be harmless to lacewing predators, as indicated by the results obtained in organic olive groves in Portugal. In agreement with these observations, in laboratory studies, Medina et al. [42] showed that the oviposition of *Ch. carnea* was not affected, and even increased on olive twigs treated with kaolinite (Surround WP) and, in line with the results of Pascual et al. [43], its application does not seem to be a problem for these oophagous predators. Likewise, Bengochea et al. [41] reported that kaolin appeared to be harmless or only slightly harmful to the green lacewings, indicating that its use might be considered as an alternative chemical product for controlling olive pests in the contexts of both organic farming and integrated olive pest management programs. However, studies on pests of other fruit crops indicate that its effectiveness seems to be highly variable depending on the target pest, even being counterproductive for certain pests [44].

Organic farming methods mitigate the ecological damage caused by aggressive agronomic management practices, which have negatively affected evenness [45]. Since these practices are often at the origin of ecological imbalances that result in an uncontrolled increase in pests, organic management practices have been considered to compensate for these alterations, promoting evenness in the complex of natural enemies [45]. These statements are corroborated by verifying the notable increase in predation rates in olive groves under organic management, exemplified in the impact on the increase in fruit recovery that was approximately 25% higher in organic olive groves compared to IPM olive groves. This corresponds to a recovery rate of between 70% and 90% of the initially infested fruit in organic olive groves. Regarding IPM crops, organic management allows for correcting the imbalances generated by environmentally aggressive practices derived from the use of synthetic pesticides and reestablishing the natural balance. Apart from the benefits due to the absence of pesticide residues in the oil, organic management allows for higher agricultural yields, reducing losses by approximately 40% and 60% in organic olive groves in southern Spain and Portugal, respectively. In comparative studies on organic and conventional management in Greece, authors such as Berg et al. [46] indicate the greater sustainability of organic management from an environmental and financial point of view, compared with conventional cultivation, resulting in healthier and better-quality olive oils.

5. Conclusions

The comparative study of the effect of entomophagous predation on *Prays oleae* in olive groves in southern Spain and Portugal has revealed great differences depending on the farming system (organic or IPM). Among the parameters considered, these differences include the population density of the pest, which in both countries has turned out to be significantly higher in organic olive groves.

Significant differences among both farming systems were also found in the rates of oophagous predation, being equally high in organic olive groves, which allowed the suppression of approximately 86% of the original egg population, ultimately resulting in the recovery of 76–80% of the fruits initially infested. In contrast, in IPM olive groves, recovery rates were detected as only 54%, approximately.

The differences in the predatory potential of lacewings imply estimated mean fruit drop rates of 28% in olive groves with IPM management and approximately half (14.5%) of that in olive groves with organic management.

Author Contributions: Conceptualization, R.G.-R. (Ramón González-Ruiz) and R.G.-R. (Roberto García-Ruiz); methodology, R.G.-R. (Ramón González-Ruiz) and J.A.G.-G.; software, J.A.G.-G.; validation, R.G.-R. (Ramón González-Ruiz) and J.A.G.-G.; formal analysis, R.G.-R. (Ramón González-Ruiz); investigation, R.G.-R. (Ramón González-Ruiz); J.A.G.-G., J.M.H., V.R., S.B. and Muñoz-Rojas, J.; resources, R.G.-R. (Roberto García-Ruiz), J.A.G.-G., J.M.H., V.R., S.B. and J.M.-R.; data curation, R.G.-R. (Ramón González-Ruiz); J.A.G.-G., J.M.H. and J.M.-R.; writing—original draft preparation, R.G.-R. (Ramón González-Ruiz); writing—review and editing J.A.G.-G., J.M.-R. and R.G.-R. (Roberto García-Ruiz); visualization, R.G.-R. (Ramón González-Ruiz) and J.A.G.-G.; supervision, R.G.-R. (Ramón González-Ruiz); project administration, R.G.-R. (Roberto García-Ruiz); funding acquisition, R.G.-R. (Roberto García-Ruiz) All authors have read and agreed to the published version of the manuscript.

Funding: This research was funded by the PRIMA-H2020 project SUSTAINOLIVE, grant number no1811. The authors from Portugal also acknowledge funding from the FCT (Fundación de Ciencia y Tecnología—Portugal), to the CHANGE LAB and the Mediterranean Institute for Agriculture, Environment and Development (MED) at the University of Évora, funded under project UIDB/05183/2020. In addition, JMH received funding from the FCT through the project PTDC/BIA-CBI/1365/2020 and the contract IF/01053/2015, and from the Spanish Government through a María Zambrano Grant.

Data Availability Statement: Not applicable.

Acknowledgments: The authors would like to thank the farmers who facilitated the sampling and provided information on the olive groves and the management practices.

Conflicts of Interest: The authors declare no conflict of interest.

References

1. Arambourg, Y. *Traité d'entomologie oléicole*; Conseil oléicole International: Madrid, Spain, 1986.
2. Ramos, P.; Campos, M.; Ramos, J.M. Long-term study on the evaluation of yield and economic losses caused by *Prays oleae* Bern. in the olive crop of Granada (southern Spain). *Crop Prot.* **1998**, *17*, 645–647. [CrossRef]
3. Álvarez, H.A.; Jiménez-Muñoz, R.; Morente, M.; Campos, M.; Ruano, F. Ground cover presence in organic olive orchards affects the interaction of natural enemies against *Prays oleae*, promoting an effective egg predation. *Agric. Ecosyst. Environ.* **2021**, *315*, 107441. [CrossRef]
4. Vilar, J.; Pereira, J.E. *La Olivicultura Internacional Difusión Histórica, Análisis Estratégico y Visión Descriptiva*; Fundación Caja Rural de Jaén: La Carolina, Spain, 2018. Available online: https://www.researchgate.net/publication/326070772_LA_OLIVICULTURA_INTERNACIONAL_DIFUSION_HISTORICA_ANALISIS_ESTRATEGICO_Y_VISION_DESCRIPTIVA (accessed on 5 September 2022).
5. Rosales, R.; Garrido, D.; Ramos, P.; Ramos, J.M. Ethylene can reduce *Prays oleae* attack in olive trees. *Crop Prot.* **2006**, *25*, 140–143. [CrossRef]
6. Ramos, P.; González, R.; Ramos, J.M. Alternativas naturales al uso de plaguicidas contra la polilla del olivo (*Prays oleae* Bern.). *Oleae* **2003**, *28*, 20–23.
7. Bento, A.; Torres, L.; Lopes, J.; Pereira, J.A. Avaliação de prejuízos causados pela traça da oliveira, Prays oleae (Bern.) em Trás-os-Montes. 2001. Available online: http://hdl.handle.net/10198/850 (accessed on 5 September 2022).
8. Arambourg, Y. Caractéristiques du Peuplement Entomologique de L'olivier Dans le Sahel de Sfax. 1964. Available online: https://www.worldcat.org/title/caracteristiques-du-peuplement-entomologique-de-lolivier-dans-le-sahel-de-sfax/oclc/459406956 (accessed on 5 September 2022).
9. Campos, M.; Ramos, P. Some relationships between the number of *Prays oleae* eggs laid on olive fruits and their predation by *Chrysoperla carnea*. In *Integrated Pesticide Control Olive-Groves Proceedings of the CEC/FAO/IOBC International Joint Meeting, Pisa Italy, 3–6 April 1984*; Balkema for CEC: Rotterdam, The Netherlands, 1985; pp. 237–241.
10. Bento, A.; Lopes, J.; Torres, L.; Passos-Carvalho, P. Biological control of *Prays oleae* (Bern.) by chrysopids in Trás-os-Montes region (Northeastern Portugal). In Proceedings of the III International Symposium on Olive Growing, Chania, Greece, 22–26 September 1997; Volume 474, pp. 535–540.
11. Corrales, N.; Campos, M. Populations, longevity, mortality and fecundity of *Chrysoperla carnea* (Neuroptera, Chrysopidae) from olive-orchards with different agricultural management systems. *Chemosphere* **2004**, *57*, 1613–1619. [CrossRef]
12. González-Ruiz, R.; Al-Asaad, S.; Bozsik, A. Influencia de las masas forestales en la diversidad y abundancia de los crisópidos (Neur.:" Chrysopidae") del olivar. *Cuad. Soc. Esp. Cienc. For.* **2008**, *26*, 33–38. Available online: https://dialnet.unirioja.es/descarga/articulo/4245523.pdf (accessed on 6 September 2022).
13. Nave, A.; Gonçalves, F.; Crespí, A.L.; Campos, M.; Torres, L. Evaluation of native plant flower characteristics for conservation biological control of *Prays oleae*. *Bull. Entomol. Res.* **2016**, *106*, 249–257. [CrossRef]
14. Mahzoum, A.M.; Villa, M.; Benhadi-Marín, J.; Pereira, J.A. Functional response of *Chrysoperla carnea* (Neuroptera: Chrysopidae) larvae on *Saissetia oleae* (Olivier) (Hemiptera: Coccidae): Implications for biological control. *Agronomy* **2020**, *10*, 1511. [CrossRef]
15. Principi, M.M.; Canard, M. Feeding habits [Chrysopidae]. Series entomologica. *Springer* **1984**, *Volume 27*, 76–92.
16. Ridgway, R.L.; Morrison, R.K.; Badgley, M. Mass rearing a green lacewing. *J. Econ. Entomol.* **1970**, *63*, 834–836. [CrossRef]
17. Koczor, S.; Knudsen, G.K.; Hatleli, L.; Szentkirályi, F.; Tóth, M. Manipulation of oviposition and overwintering site choice of common green lacewings with synthetic lure (Neuroptera: Chrysopidae). *J. Appl. Entomol.* **2015**, *139*, 201–206. [CrossRef]
18. Grafton-Cardwell, E.E.; Hoy, M.A. Intraspecific Variability in Response to Pesticides in the Common Green Lacewing, *Chrysoperla carnea* (Stephens)(Neuroptera: Chrysopidae). *Hilgardia* **1985**, *53*, 1–32. [CrossRef]
19. Tauber, M.J.; Tauber, C.A.; Daane, K.M.; Hagen, K.S. Commercialization of predators: Recent lessons from green lacewings (Neuroptera: Chrysopidae: Chrosoperla). *Am. Entomol.* **2000**, *46*, 26–38. [CrossRef]
20. Thierry, D.; Cloupeau, R.; Jarry, M.; Canard, M. Discrimination of the West-Palaearctic Chrysoperla Steinmann species of the *carnea* Stephens group by means of claw morphology (Neuroptera, Chrysopidae). *Acta Zool. Fenn.* **1998**, *209*, 255–262.

21. Henry, C.S.; Brooks, S.J.; Thierry, D.; Duelli, P.; Johnson, J.B. The common green lacewing (*Chrysoperla carnea* s. lat.) and the sibling species problem. In *Lacewings in the Crop Environment*; Cambridge University Press: Cambridge, UK, 2001; pp. 29–42. [CrossRef]
22. Henry, C.S.; Brooks, S.J.; Duelli, P.; Johnson, J.B. Discovering the true *Chrysoperla carnea* (Insecta: Neuroptera: Chrysopidae) using song analysis, morphology, and ecology. *Ann. Entomol. Soc. Am.* **2002**, *95*, 172–191. [CrossRef]
23. Henry, C.S.; Brooks, S.J.; Duelli, P.; Johnson, J.B. A lacewing with the wanderlust: The European song species 'Maltese', *Chrysoperla agilis*, sp. n., of the *carnea* group of Chrysoperla (Neuroptera: Chrysopidae). *Syst. Entomol.* **2003**, *28*, 131–148. [CrossRef]
24. Bozsik, A.; González-Ruíz, R.; Lara, B.H. Distribution of the *Chysoperla carnea* complex in southern Spain (Neuroptera: Chrysopidae). *An. Univ. Oradea Fasc. Protecția Mediu.* **2009**, *14*, 60–65. Available online: https://protmed.uoradea.ro/facultate/anale/protectia_mediului/2009/agr/11.Bozsikv%20Andras%202.pdf (accessed on 7 September 2022).
25. Bollero, A.L.; Moya, J.H.; Macías, V.V.; Mohedano, D.P.; Moya, J. *Introducción al Olivar Ecológico en Andalucía*; Instituto de Investigación y Formación Agraria y Pesquera, Junta de Andalucía: Seville, Spain, 2017.
26. Ehi-Eromosele, C.; Nwinyi, O.C.; Ajani, O.O. Integrated Pest Management. In *Weed and Pest Control—Conventional and New Challenges*; IntechOpen: London, UK, 2013. [CrossRef]
27. Ehler, L.E. Integrated pest management (IPM): Definition, historical development and implementation, and the other IPM. *Pest Manag. Sci.* **2006**, *62*, 787–789. [CrossRef]
28. Ponzio, C.; Gangatharan, R.; Neri, D. Organic and biodynamic agriculture: A review in relation to sustainability. *Int. J. Plant Soil Sci.* **2013**, *2*, 95–110. [CrossRef]
29. Besson, Y. Une histoire d'exigences: Philosophie et agrobiologie. L'actualité de la pensée des fondateurs de l'agriculture biologique pour son développement contemporain. *Innov. Agron.* **2009**, *4*, 329–362.
30. Gegner, L.; Kuepper, G. Organic Crop Production Overview. *Attra Publ. MP* **2004**, *170*, 1–28.
31. Bengtsson, J.; Ahnström, J.; Weibull, A. The effects of organic agriculture on biodiversity and abundance: A meta-analysis. *J. Appl. Ecol.* **2005**, *42*, 261–269. [CrossRef]
32. Gómez-Guzmán, J.A.; García-Marín, F.J.; Sáinz-Pérez, M.; González-Ruiz, R. Behavioural resistance in insects: Its potential use as bio indicator of organic agriculture. In *IOP Conference Series: Earth and Environmental Science*; IOP Publishing: Bristol, UK, 2017; Volume 95, p. 042038.
33. Gómez-Guzmán, J.A.; Sainz-Pérez, M.; González-Ruiz, R. Monitoring and Inference of Behavioral Resistance in Beneficial Insects to Insecticides in Two Pest Control Systems: IPM and Organic. *Agronomy* **2022**, *12*, 538. [CrossRef]
34. Scherber, C.; Milcu, A.; Partsch, S.; Scheu, S.; Weisser, W.W. The effects of plant diversity and insect herbivory on performance of individual plant species in experimental grassland. *J. Ecol.* **2006**, *94*, 922–931. [CrossRef]
35. Popp, J.; Hantos, K. The impact of crop protection on agricultural production. *Stud. Agric. Econ.* **2011**, *113*, 47–66. [CrossRef]
36. Colbrant, P.; Fabre, P. Stades Reperes de l'olivier. *L'Olivier. Invuflec*, Paris 1975, 24–25. Available online: http://afidol.org/wp-content/uploads/2016/04/stades_pheno.pdf (accessed on 7 September 2022).
37. Ramos, P.; Ramos, J.M. Veinte años de observaciones sobre la depredación oófaga en *Prays oleae* Bern. Granada (España), 1970–1989. *Boletín Sanid. Veg. Plagas* **1990**, *16*, 119–127.
38. Razali, N.M.; Wah, Y.B. Power comparisons of shapiro-wilk, kolmogorov-smirnov, lilliefors and anderson-darling tests. *J. Stat. Model. Anal.* **2011**, *2*, 21–33.
39. Bozsik, A.; González-Ruiz, R. First data on the sibling species of the common green lacewings in Spain (Neuroptera: Chrysopidae): (The taxonomic status of the most important cryptic species of *Chrysoperla carnea* complex in Spain). In Proceedings of the 4th International Plant Protection Symposium at Debrecen University and 11th Trans-Tisza Plant Protection Forum, Debrecen, Hungary, 18–19 October 2006; Debreceni Egyetem, Agrártudományi Centrum, Mezögazdaságtudományi Kar: Debrecen, Hungary, 2006; pp. 3–11.
40. Pascual, S.; Cobos, G.; Seris, E.; González-Nuñez, M. Field assessment of kaolin as a pest control tool in an olive grove in Madrid. *IOBC/WPRS Bull.* **2010**, *53*, 109–113.
41. Bengochea, P.; Saelices, R.; Amor, F.; Adán, Á.; Budia, F.; del Estal, P.; Viñuela, E.; Medina, P. Non-target effects of kaolin and coppers applied on olive trees for the predatory lacewing *Chrysoperla carnea*. *Biocontrol Sci. Technol.* **2014**, *24*, 625–640. [CrossRef]
42. Medina, P.; Budia, F.; Contreras, G.; Del Estal, P.; González-Núñez, M.; García, M.; Viñuela, E.; Adán, A. Comportamiento reproductor del depredador *Chrysoperla carnea* (Stephens)(Neuroptera: Chrysopidae) y el parasitoide *Psyttalia concolor* (Szèpligeti) (Hymenoptera: Braconidae) en superficies cubiertas con caolín. In *V Congreso Nacional de Entomología Aplicada*; SEdE Aplicada: Cartagena, Spain, 2007; pp. 22–26.
43. Pascual, S.; Cobos, G.; Seris, E.; González-Núñez, M. Effects of processed kaolin on pests and non-target arthropods in a Spanish olive grove. *J. Pest Sci.* **2010**, *83*, 121–133. [CrossRef]
44. Marko, V.; Blommers, L.; Bogya, S.; Helsen, H. Kaolin particle films suppress many apple pests, disrupt natural enemies and promote woolly apple aphid. *J. Appl. Entomol.* **2008**, *132*, 26–35. [CrossRef]
45. Crowder, D.W.; Northfield, T.D.; Strand, M.R.; Snyder, W.E. Organic agriculture promotes evenness and natural pest control. *Nature* **2010**, *466*, 109–112. [CrossRef] [PubMed]
46. Berg, H.; Maneas, G.; Salguero Engström, A. A comparison between organic and conventional olive farming in Messenia, Greece. *Horticulturae* **2018**, *4*, 15. [CrossRef]

Article

Assessment of Thrips Diversity Associated with Two Olive Varieties (Chemlal & Sigoise), in Northeast Algeria

Randa Mahmoudi [1,*], Malik Laamari [1] and Arturo Goldarazena [2]

[1] LATPPAM Research Laboratory, Department of Agriculture, University of Batna 1, Batna 05000, Algeria
[2] National Museum of Natural Sciences, National Reference Laboratory for Nematodes and Arthropods, Department of Biodiversity and Evolutionary Ecology, Calle Serrano 115, 28006 Madrid, Spain
* Correspondence: randa.mahmoudi@univ-batna.dz

Abstract: In this study, the diversity of thrips (Insecta: Thysanoptera) on two varieties of olive trees (Chemlal and Sigoise) in northeast Algeria (Province of Batna), was evaluated for 3 years (2019–2021). In addition, the fluctuations in the numbers of phytophagous thrips were estimated according to the varieties phenological stages. A total of 19 species are identified and the olive thrips (*Liothrips oleae*) have just been reported for the first time in Algeria. Only 5 females of this species were collected in May 2021 on the Sigoise variety at the fruit-setting stage. *Haplothrips tritici*(17.25%), *Frankliniella occidentalis* (16.29%) and *Thrips tabaci* (16.29%) are the most present. It is noticed that the thrips were present on the olive tree only in spring (April to May), when the average monthly temperatures are between 10–26 °C, but linear regression analyses were not confirmed that temperature explain the variation in thrips numbers, which may be due to other climatic factors such as the rainfall, while olive varieties and phenological stages are affecting the population of thrips, their number was higher on the Sigoise variety, especially at flowering stage in the case of *H. tritici* and *F. occidentalis* while *T. tabaci* was most noticeable at the fruit growth stage. The number of this species was relatively low, just until the inflorescence stage, where thrips start to appear in Sigoise before Chemlal.

Keywords: olive tree; phytophagous; Thysanoptera

Citation: Mahmoudi, R.; Laamari, M.; Goldarazena, A. Assessment of Thrips Diversity Associated with Two Olive Varieties (Chemlal & Sigoise), in Northeast Algeria. *Horticulturae* **2023**, *9*, 107. https://doi.org/10.3390/horticulturae9010107

Academic Editors: Małgorzata Tartanus and Eligio Malusà

Received: 12 December 2022
Revised: 9 January 2023
Accepted: 10 January 2023
Published: 12 January 2023

Copyright: © 2023 by the authors. Licensee MDPI, Basel, Switzerland. This article is an open access article distributed under the terms and conditions of the Creative Commons Attribution (CC BY) license (https://creativecommons.org/licenses/by/4.0/).

1. Introduction

The cultivation of the olive tree (*Olea europea* L.) occupies an important place in the economy of Mediterranean countries, particularly in North Africa, where it plays a very important socio-economic and environmental role [1].unfortunately, throughout the world, this crop is subject to several pests, which negatively affect yields in quantity and quality and annual losses are estimated at more than 15% [2].

Among these pests, phytophagous thrips occupy a very important place. According to Marullo and Vono [3], food bites by can be observed on flower buds and young leaves. On the leaves, these attacks cause necrosis, desiccation, and deformation. On fruit, these stains cause deformation, drying out, and premature drop. These various types of damage lead above all to a reduction in oil yield.

Despite the extent of their damage to olive trees, thrips remain among the least studied groups of insects, particularly in the Maghreb countries. Among the works relating to olive thrips in countries around the Mediterranean, there are especially those of Rei et al. [4] in Portugal, Canale et al. [5] in Italy, Agamy et al. [6] in Egypt, and more recently, Halimi et al. [7] in Algeria. The olive thrips *Liothrips oleae* was already reported in Spain during the nineteen century also in France [8,9], in the Maltese islands [10], and in Italy [3,11], in Algeria it is not yet mentioned. In addition to these records in Portugal, Spain, Greece, and Italy, this thrips is noted more also in Poland, Ethiopia, and Yemen [12,13].

The olive grove in our study consists of two varieties (Chemlal and Sigoise) randomly distributed in the study orchard, of which Chemlal, considered a hardy and late variety, originating from Kabylia, occupies 40% of the Algerian olive orchard with high productivity intended for the production of oil, with a productivity of 14 to 18 L/quintal [14], on the other hand, Sigoise considered as a seasonal variety, early and tolerant to salt water, their origin is Plane de Sig (Mascara), occupies 25% of the Algerian olive orchard, intended for the production of table olives and oil with average productivity [15].

Polyphagous insects are classified as phytophagous, mycophagous, and predatory, with phytophagous thrips feeding on flowers, fruit, mature leaves, or flower buds [16,17], consequently, so the characteristics and phenological stages of olive varieties can be contributed to their abundant numbers; which can develop in parallel with the different phenological stages of the olive tree; from the hatching of the axillary buds and the appearance of new terminal shoots, in spring; until the flowering as the spring temperature becomes milder, then fruiting and ripening [18].

The principles objective of this study was to evaluate the diversity of thrips associated with olive trees in northeastern Algeria (province of Batna), as the species may be harmful to olive trees, and to study fluctuations in the numbers of thrips adult, according to temperatures and olive varieties phenological stages, during the 2019/2021 period.

2. Materials and Methods

2.1. Study Site

This study was carried out in an olive grove located in the province of Batna (northeastern Algeria) (6°24′54.72″ E, 35°42′30.24″ N, 875 m) and which is characterized by asemi-arid climate, where cereal growing is the main crop in this region. This olive grove was installed in 2002 and is occupied by 1000 olive trees belonging to the Chemlal and Sigoise varieties. The only standard agronomic practices in the orchard aredeep plowing, drip irrigation system, and tree pruning, without including insecticide application, which is considered an ecological cultivation area. Also, the study orchard is surrounded by a few apple trees and many olive groves.

2.2. Diversity of Thrips Communities on Olive Trees

During the period from 2019 to 2021, the olive grove prospected twice a month and the sampling was carried out according to the method proposed by De Borbon [19]. At each survey, the trees were evaluated by randomly, and the canopy lower section of 10 trees per variety (each tree in 4 directions: east, west, north, and south) was beaten by hand on a white plastic tray. Captured specimens were stored in microtubes containing 70% ethanol. According to Reynaud et al. [20], this solution maintains the flexibility of these thrips. Preserved specimens are transferred to NaOH (5%) solution between 2 h (light individuals) and 2 days (dark individuals). Then, the samples were mounted on slides in Hoyer's medium. Voucher specimens were mounted in Canada balsam. Identification of adults was made using the keys provided by zur Strassen and Moritz [21,22]. The main microscopic characters retained for this identification are the number of antennal segments, the shape, and the number of sensory cones, the wing venation, and the number and size of setae on the pronotum [23]. Voucher specimens were deposited in the insect collection of LATPPAM Research Laboratory, University of Batna, and in the National Reference Laboratory for Nematodes and Arthropods, National Museum of Natural Sciences, Madrid.

2.3. Statistical Analyzes

Statistical analyses were carried out to determine the variations in the number of thrips species according to varieties and phenological stages. Data were subjected to linear regression analysis, to determine the correlation between average temperatures monthly and thrips abundance, in addition to the study of variance two-way ANOVA test of highly significant differences (HSD) at $p < 0.05$, to analyze the relationship between the thrips numbers and the two independent variables, olives varieties and phenological stages, and the Scheffe test to find out which pairs of means are significant. All analyzes were performed using Microsoft Statistics SPSS version 25 [24].

3. Results

3.1. Thrips Diversity of on Olive Trees

The results revealed the presence of 19 species of thrips on olive trees during three years of the survey. Among these species, seven species have just been reported for the first time in Algeria among these, *Liothrips oleae* (Figure 1). More than four thrips species were recorded for the first time on olive trees over the world (Table 1), along with, *Liothrips leucopus* (Figure 2). Most thrips identified, were phytophagous (92.01%), and only (7.99%) were facultative predators.

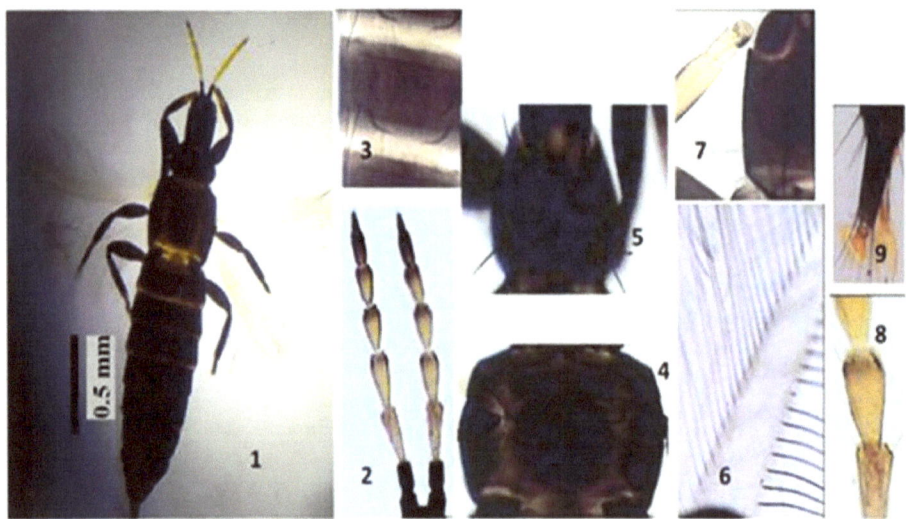

Figure 1. 1, *Liothrips oleae* female: 2, antenna; 3, abdominal tergites; 4, mesonotum and metanotum; 5, pronotum; 6, wing; 7, head; 8, antennal segments III–IV; 9, segments IX–X.

Figure 2. 1, *Liothrips leucopus*: 2, antennal segments III–IV; 3, head; 4, segments IX–X; 5, wing pronotum; 6, pronotum; 7, mesonotum and metanotum; 8, abdominal tergites.

Table 1. The different species of thrips recorded on olive trees in the province of Batna (2019–2021).

Suborder	Family	Species	Feed
Terebrantia	Thripidae	*Frankliniella occidentalis* Pergande, 1895	Phyt.
		Thrips tabaci Lindeman, 1889	Phyt.
		Thrips angusticeps Uzel, 1895	Phyt.
		(*) *Anaphothrips obscurus* Muller, 1776	Phyt.
		Thrips minutissimus Linnaeus, 1758	Phyt.
		(*) *Stenothrips graminum* Uzel, 1895	Phyt.
		(*) *Tenothrips frici* Uzel, 1895	Phyt.
		(*) *Dendrothrips ornatus* Jablonowski, 1894	Phyt.
	Melanthripidae	*Melanthrips fuscus* Sulzer, 1776	Phyt.
	Aeolothripidae	(*) *Aeolothrips tenuicornis* Bagnall, 1926	Fac. Pred.
		Aeolothrips intermedius Bagnall, 1934	Fac. Pred.
Tubulifera	Phlaeothripidae	(*) *Liothrips oleae* Costa, 1857	Phyt.
		(*) (**) *Liothrips leucopus* Titschack, 1958	Phyt.
		(**) *Dolicholepta micrura* Bagnall, 1914	Pred.
		Haplothrips tritici Kurdjumov, 1912	Phyt.
		Haplothrips andresi Priesner, 1931	Phyt.
		(**) *Haplothrips crassicornis* John, 1924	Phyt.
		(**) *Haplothrips distinguendus* Uzel, 1895	Phyt.
		(*) *Haplothrips aculeatus* Fabricius, 1803	Phyt.

(*): First record in Algeria, (**): First record on olive tree, Phyt.: Phytophagous, Fac. Pred.: Facultative predator.

3.2. Thrips Adults numbers on Olive Trees

313 specimens of the 19 thrips species were collected in this study; it revealed a variation in the distribution of thrips over the three years of sampling, the majority of them was in 2019 by the maximum number of thrips (155 specimens), followed by 98 and only 60 specimens in 2021 and 2020, respectively, the principal species were *H. tritici, F. occidentalis, T. tabaci,* and *Thrips angusticeps*, accounting for 17.25%, 16.29%, 16.29%, and 10.87% of the total number of thrips, respectively (Figure 3).

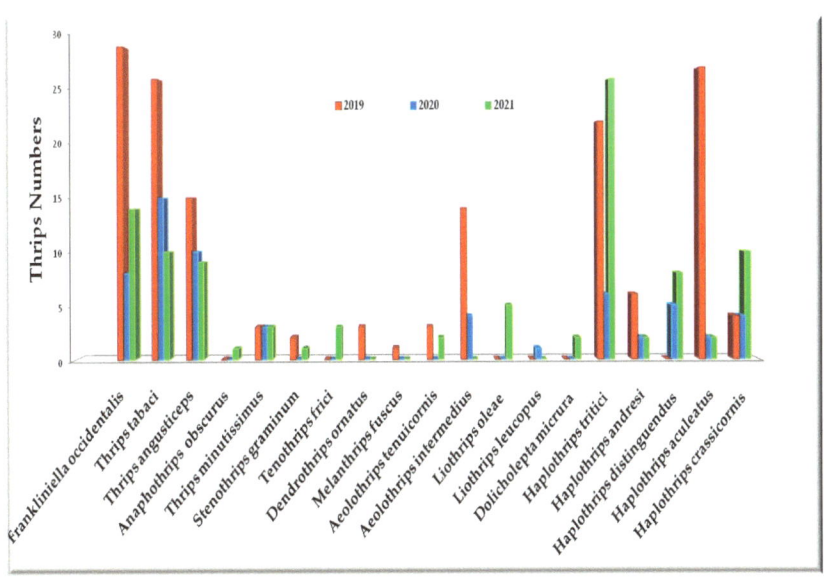

Figure 3. Temporal variation of thrips numbers in olive trees in the province of Batna (north-eastern Algeria) during the 2019/2021 period.

3.3. Abundance of Thrips Adult Numbers According to the Temperatures

During the period of the research, thrips adults on olive trees have not determined in winter (January and February). while it appears in spring (March–May), where during May they recorded the maximum of his numbers when the average monthly temperatures are between 10–26 °C. But in the summer months, their numbers fall again with the increase in temperatures until they disappear in autumn (Figure 4). Also, we have noticed that April to May 2021 shows anomalous values that are not consistent with the results for 2020 and 2019, this may be due to other climatic factors.

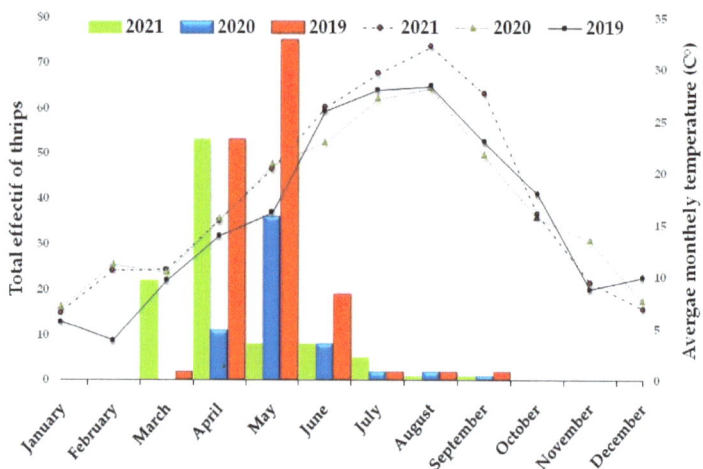

Figure 4. Temporal fluctuation of thripsadultson olive trees in Batna region according to monthly average temperatures.

Linear regression analyses (Figure 5) were not significant ($p = 0.37$) and the temperature doesn't explain a proportion of variation in the rate of thrips numbers, where the value showed a poorly positive correlation between average temperatures and thrips abundance, where the temperature doesn't contribute in the variation in thrips abundance (R-Square = 0.010).

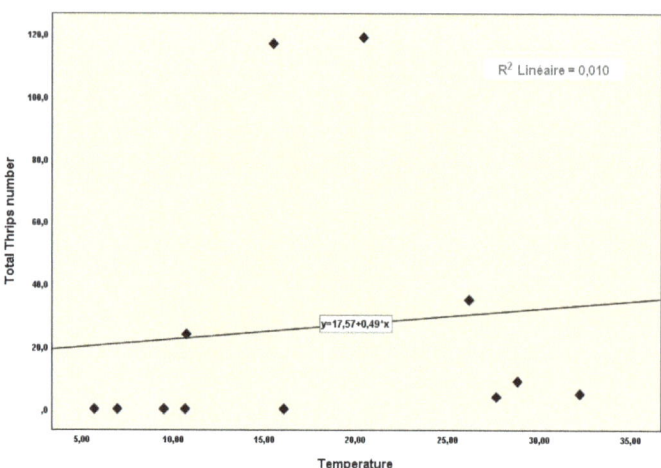

Figure 5. Thrips numbers relationship with temperatures average in Batna region.

3.4. Distribution of Thrips adults Numbers According to Olive Varieties Phenological Stages

The results obtained prove that on the Sigoise variety thrips numbers were higher than it was on the Chemlal variety. Results of statically testing for the influence of the varieties and phenological stages on the presence and abundance of thrips species showed that there was a significant effect of each factor on the total number of thrips with (F = 19.60, df = 1, $p = 0$), and (F = 11.68, df = 8, $p = 0$) successively, also specifying which stage is favorable for each variety, where the fruit growth (1st stage) was the most attractive for the Sigoise variety, while the average number of this species during the flowering stage is significantly higher than the other phenological stages in the Chemlal variety (Figure 6).

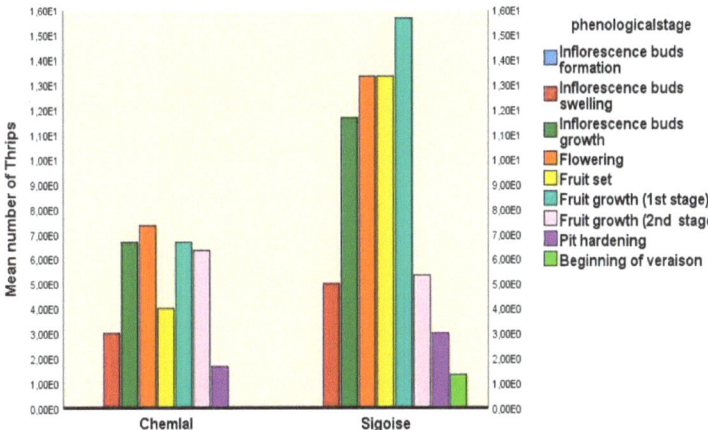

Figure 6. Mean number of Thrips adults according to olive varieties phenological stages in Batna region.

3.5. The Principal Phytophagous Species Numbers

The abundance of the principal phytophagous species varied according to olive varieties phenological stages, and the activity of these thrips was more important on Sigoise than the Chemlal variety. Scheffe test confirmed the result obtained, with a significant value(F = 9, df = 8, $p = 0$), where the average number of *H. tritici* (3.33 individuals) (Figure 7), during the Flowering stage was significantly higher than the other phenological stages in both varieties. The same result concerning the phenological stage preferably was observed with *F. occidentalis* (3.67 individuals) (Figure 8), where the flowering stage was the most attractive, while the main number of *T. tabaci* (Figure 9) was more important in the fruit growth stage with(2.66 individuals) and (F = 8.83, df = 8, $p = 0$).

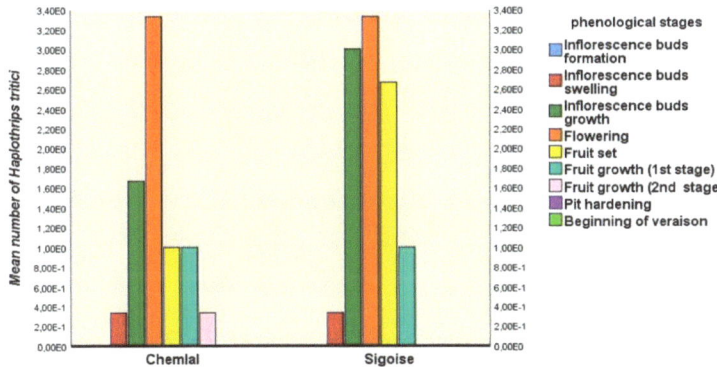

Figure 7. Mean number of *H. tritici* according to olive varieties phenological stages in Batna region.

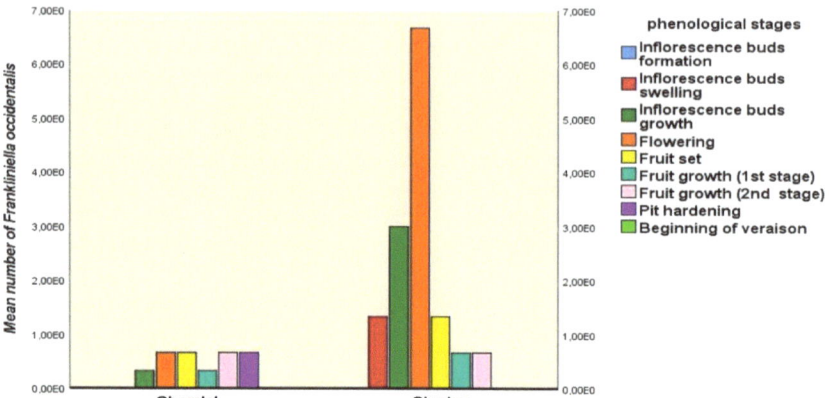

Figure 8. Mean number of *F. occidentalis* according to olive varieties phenological stages in Batna region.

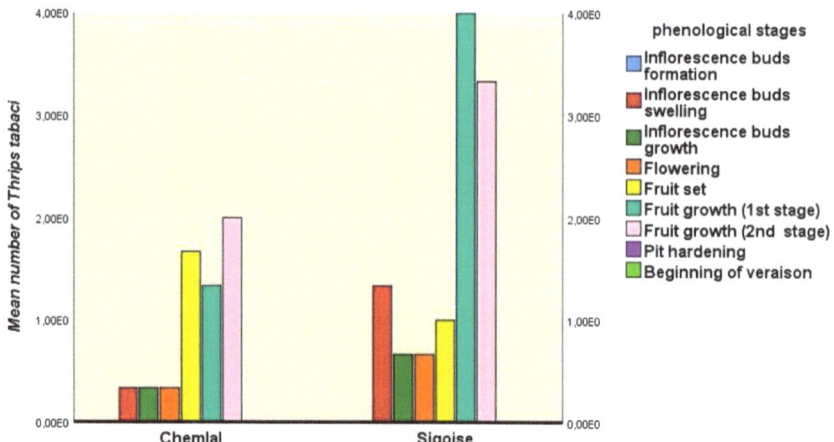

Figure 9. Mean number of *T. tabaci* according to olive varieties phenological stages in Batna region.

4. Discussion

The majority of thrips species collected on olive trees were phytophagous, more than predators; which may be due to the nature of the study area which is highly agricultural, and the regular agricultures techniques practice, such as irrigation by drop. Some of these species found in the olive trees probably do not reproduce in the olive leaves or fruits. For example, *A. obscurus, S. graminum, H. tritici*, and *H. aculeatus* breeding Gramineae (Poaceae) and *Phragmites australis*. *T. minutissimus* laid eggs on flowers of Quercus in Europe and in Spain and is frequently collected in flowers of Genista, Bellis, and Sinapis [25]. *T. angusticeps* is associated with Cruciferae and *Linum sp.* but it has been collected in Poaceae, Fabaceae, and Compositae in Mediterranean and eurosiberian Spain [25,26]. *M. fuscus* is also abundant in flowers of Brassicaceae which it probably breeds [27]. The presence of these species in the olive leaves and flowers could be possible due to the ecological cultivation of olives in the area of sampling, as no chemical treatment is applied, where the presence of many indigenous plants between the trees is common. *Liothrips leucopus* was originally described from one male collected in *Quercus ilex* in Montpellier (France) [28], however, some populations have been collected posteriorly in Andalusia (Spain) and Marocco by Dr. Richard zur Strassen. Looking carefully at these specimens kindly borrowed by The Senckenberg Museum (Frankfurt am Main, Germany) where are deposited, we observed a variation in the color of the femora between the populations collected in Spain and North Africa. The specimens collected in Andalusia (Spain) have brown femora in contrast with the North African specimens with yellow femora and tibiae. The Algerian specimen has the leg completely yellow, similar to the Maroccan specimens. Unfortunately, we do not observe more differences between the specimens of both populations to support a specific separation between both populations.

The samples taken in our study revealed greater biodiversity (16 phytophagous species and 3 predatory species), compared with all other works relating to olive thrips, such as Rei et al. [4] in Portugal where they identifted9 species of thrips (8 of them are phytophagous), also Canale et al. [5] in Italy has determined 14 phytophagous species and only 2 predatory species, while in Egypt Agamy et al. [6] they represented only 7 phytophagous species, and more recently, Halimi et al. [7] in Algeria with 9 species (7 phytophagous and 2 predatory).

Temperature is one seasonal factor that might impact thrips population dynamics [29,30]. According to Lowry et al. [29], they observed on Peanut, that *Frankliniella fusca* and *F. occidentalis* have an inverse connection between developmental time and temperature, as is typical for all insects. Where the number of degree days required to complete one generation for this thrips reported being 234 (lower threshold 10.5 °C) and 253 (6.5 °C), respectively.

According to Loomans and van Lenteren [31], temperatures between 25 and 30 °C are ideal for the development of thrips, however, during the period of the research, the average monthly temperatures for winter (January and February) were not higher than 5 °C, which may have had a negative impact on thrips development because it would not have been able to resume its activities at these temperatures. While the ambient temperatures observed during May recorded the maximum thrips numbers. In general, thrips were present on the olive tree only in spring (March–May), when the average monthly temperatures are between 10–26 °C. It appears that the May activity is more significant. Additionally, Halimi et al. [7] confirmed that the spring activity of the thrips *Neohydatothrips amygdali* on olive trees is more important, recorded in May (spring) more than in October (autumn).

Also, the anomalous values that are noticed that April to May 2021; which are not consistent with the results for 2020 and 2019, maybe due to other climatic factors, such as the rains, because, when we checked the climatic data, we noticed a difference in the rainfall during these two months, where it is in decreasing (from 53 mm to 29.6 mm in 2019, and from 29.9 mm to 22.5 mm in 2020), while in 2021 it is increasing (from 32.2 mm to 47 mm), According to Bournier [30], the temperature and the hygrometry act in parallel, on the effect in thrips numbers, of which the heavy rains are responsible for the destruction of the majority of their populations.

Actually, the fluctuation of the seasonal thrips population is influenced by various environmental factors including climate, host-plant variety, topography, soil type, and management regimes [32]. According to Etebari et al. [32], the mulberry thrips (*Pseudodendrothrips mori*) show some host preference in their feeding activity, whereas the Kenmuchi variety is preferred by thrips, being that has more proteins which are important nutrients for thrips development in comparison to the other mulberries varieties. Thierry et al. [33] add also that the selection of plants or varieties by phytophagous insects may depend in part on the physical characteristics of feeding and oviposition sites. This includes leaf size, shape, color, thickness, the density of trichomes and stomata, the presence of minerals such as epidermal wax crystals or silica, also the texture and relief of plant surface.

Moreover, as the two varieties are exposed to the same environmental conditions and benefit from the same cultivation techniques, differences in thrips numbers can be only attributed to intrinsic factors, with their morphological differences and the abundance of some primary and secondary metabolites being related. In fact, Mendel & Sebai [15], confirmed the variation in morphological characteristics of each olive variety. They showed that Sigoise is an early maturing variety with medium vigor, long leaves, few flowers, and ovoid fruits, while Chemlal is a late maturing variety with high vigor, average-sized leaves, medium number of flowers, and elongated fruits. Consequently, the early maturity of the variety also contributed to the attractiveness of thrips, because it is noticed that, the Sigoise variety which was the most preferred, reached its early maturity about two weeks before the Chemlal variety. This was confirmed by the results of Voorrips et al. [34], which show that early plant development leads to a larger thrips population and more severe damage later in the season.

During the period of the early phenological stages of both varieties, the number of this species was relatively low, according to Halimi et al. [7]. This may be due to the absence of flowers and fruit on the olive trees, because just until the inflorescence buds swelling and growth, where thrips start to appear in Sigoise before Chemlal. However, after the bloom period, their numbers increased considerably. They registered their major presence by maximum numbers in April and May in both varieties. This maximum number coincides with the flowering, fruit set, and growth stages. Our results are in agreement with those obtained by Halimi et al. [7], who mentioned that in the olive tree the activity of *N. amygdali* starts only after fruit formation (fruit set), which occurs before the end of April and the beginning of May. Similarly, *F. occidentalis* and *H. tritici* were present in a single peak in April and early May. This peak coincides with the flowering stage.

5. Conclusions

The results revealed the presence of 19 species of thrips on olive trees in northeast Algeria (Province of Batna), during three years of the survey (2019–2021), among these species, seven species have just been reported for the first time in Algeria among these, *L. oleae*, and More than four thrips species were recorded for the first time on olive trees over the world, along with, *L. leucopus*. The majority of thrips species collected on olive trees were phytophagous, more than predators.

Although the temperature is one seasonal factor that might impact thrips population dynamics, also we noticed that ambient temperatures observed in spring (March–May) recorded the maximum thrips numbers, when the average monthly temperatures are between 10–26 °C. however, Linear regression analyses were not significant and the temperature doesn't explain a proportion of variation in the rate of thrips numbers, which may be due to other climatic factors; because, according to climatic data during the study months, we noticed an inverse relationship between thrips numbers and the rainfall, in which that heavy rains are responsible for the destruction of the majority of their populations.

While olive varieties and phenological stages are important factors affecting the population dynamics of *H. tritici*, *F. occidentalis*, and *T. tabaci*, the activity of these thrips was more important on Sigoise than the Chemlal variety, where the flowering stage was the most attractive of *H. tritici* and *F. occidentalis*. Despite this stage did not attract *T. tabaci*, the fruit growth stage caused its high abundance.

During the period of the early phenological stages of both varieties, the number of this species was relatively low, which may be due to the absence of flowers and fruit on the olive trees, because just until the inflorescence buds swelling and growth, where thrips start to appear in Sigoise before Chemlal. However, after the bloom period, their numbers increased considerably. They registered their major presence by maximum numbers in April and May in both varieties. This maximum number coincides with the flowering, fruit set, and growth stages.

Author Contributions: All authors contributed to the revision of the manuscript. R.M. worked on all experiments in the laboratory, the collection trips, the data analyses, and the writing of the manuscript. M.L. worked on the designed and plan of the study, supervised the experiments, and contributed to writing the manuscript. A.G. worked on the identification of thrips species. All authors have read and agreed to the published version of the manuscript.

Funding: This research received no external funding.

Institutional Review Board Statement: The insect collections were made between 2019 and 2021 through the certificate issued by University of Batna 1 with the permission to collect specimens with scientific research, Thrips were collected on private property and permission was received from the landowners before sampling procedures.

Data Availability Statement: The datasets supporting the conclusions of this article are included within the article and its additional file.

Acknowledgments: We thank the owners of the orchard who permitted for the collection of biological material. We thank Andrea Vesmanis (The Senckenberg Museum, Frankfurt) for sending slides to certified *Liothrips leucopus* and Laurence Mound for advice to search the holotype of *L. leucopus*.

Conflicts of Interest: The authors declare no conflict of interest.

References

1. Bensemmane, A. Développons Le Secteur de l'Huile d'Olive En Algérie. *Filaha Innove* **2009**, *4*, 7.
2. Bueno, A.M.; Jones, O. Alternative Methods for Controlling the Olive Fly, Bactrocera Oleae, Involving Semiochemicals. *IOBC WPRS Bull.* **2002**, *25*, 147–156.
3. Marullo, R.; Vono, G. Forti Attacchi Di Liothrips Oleae Su Olivo in Calabria. *Inf. Agrar.* **2017**, *36*, 51–55.
4. Rei, F.T.; Mateus, C.; Torres, L. Thrips in *Oleae europaea* L.: Organic versus Conventional Production. *Acta Hortic.* **2011**, *924*, 151–156. [CrossRef]
5. Canale, A.; Conti, B.; Petacchi, R.; Rizzi, I. Thysanoptera Collected in an Onlive-Growing Area o the Northern Tuscany (Italy). *Entomol. Probl.* **2003**, *33*, 105–110.
6. Agamy, E.A.; El-Husseini, M.M.; El-Sebaey, I.I.; Wafy, M.E. The Egyptian Thripid Species in Olive Groves at Ismaialia, Egypt. *Egypt. Acad. J. Biol. Sci. A Entomol.* **2017**, *10*, 1–17. [CrossRef]
7. Halimi, C.; Laamari, M.; Goldarazena, A. A Preliminary Survey of Olive Grove in Biskra (Southeast Algeria) Reveals a High Diversity of Thrips and New Records. *Insects* **2022**, *13*, 397. [CrossRef]
8. Bejarano-Alcázar, J.D.; Rodríguez-Jurado, J.M.; Durán-Álvaro, M.; Ruiz-Torres, M.; Herrera-Mármol, M. Unidad Didáctica 5. Control de Enfermedades y Plagas En Producción Integrada Del Olivar. In *Producción Integrada de Olivar*; Instituto de Investigación y Formación Agraria y Pesquera: Junta de Andalucía, Spain, 2011.
9. Tamburin, F. *Mémoire Sur le Thrips Olivarius, (Thrips de l'Olivier) et Sur Les Moyens de Prévenir les Ravages de Cet Insecte, Draguignan*; Imprimerie de il Bernard, Près la Paroisse: Draguignan, France, 1842.
10. Haber, G.; Mifsud, D. Pests and Diseases Associated with Olive Trees in the Maltese Islands (Central Mediterranean). *Cent. Medit. Nat.* **2007**, *4*, 143–161.
11. Vono, G.; Bonsignore, C.P.; Gullo, G.; Marullo, R. Olive Production Threatened by a Resurgent Pest Liothrips Oleae (Costa, 1857) (Thysanoptera: Phlaeothripidae) in Southern Italy. *Insects* **2020**, *11*, 887. [CrossRef]
12. Kucharczyk, H.; Zawirska, I. On the Occurrence of Thysanoptera in Poland. In *Thrips and Tospoviruses, Proceedings on the 7th International Symposium on Thysanoptera, Reggio Calabria, Italy, 7–11 July 2001*; CSIRO Entomology: Clayton South, Australia, 2001; pp. 341–344.
13. Morison, G.D. Thysanoptera from South-West Arabia and Ethiopia. *J. Proc. Linn. Soc.* **1958**, *43*, 587–598. [CrossRef]
14. Lamani, O.; Ilbert, H. Spécificités de l'Oléiculture en Montagne (Région Kabyle en Algérie): Pratiques Culturales et Enjeux de la Politique Oléicole Publique. *L'Oléiculture Au Maroc de la Préhistoire à Nos Jours: Pratiques, Diversité, Adaptation, Usages, Commerce et Politiques*; CIHEAM: Montpellier, France, 2016; pp. 149–159.
15. Mendil, M.; Sebai, A. *Catalogue des Variétés Algériennes de l'Olivier. Ministere de l'Agriculture et Du Développement Rural*; ITAF Alger: Algeria, 2006.
16. Kirk, W.D.J. Pollen-feeding in Thrips (Insecta: Thysanoptera). *J. Zool.* **1984**, *204*, 107–117. [CrossRef]
17. Lewis, T. *Thrips, Their Biology, Ecology and Economic Importance*; Academic Press: London, UK; New York, NY, USA, 1973; p. 349.
18. Loussert, R.; Brousse, G. *L'Olivier. Techniques Agricoles et Production Méditerranéennes*; Maisonneuve et Larose: Paris, France, 1978.
19. De Borbon, C.M. Desertathrips Chuquiraga Gen. et Sp. n. (Thysanoptera, Thripidae) from Argentina. *Zootaxa* **2008**, *1751*, 25–34. [CrossRef]
20. Reynaud, P.; Balmes, V.; Pizzol, J. Thrips Hawaiiensis (Morgan, 1913) (Thysanoptera: Thripidae), an Asian Pest Thrips Now Established in Europe. *EPPO Bull.* **2008**, *38*, 155–160. [CrossRef]
21. Zur-Strassen, R. Die Terebranten Thysanopteren Europas und des Mittelmeer-Gebietes. *Die Tierwelt Dtschl.* **2003**, *74*, 1–277.
22. Moritz, G. Pictorial Key to the Economically Important Species of Thysanoptera in Central Europe. *EPPO Bull.* **1994**, *24*, 181–208. [CrossRef]
23. Mound, L.A. Species of the Genus Thrips (Thysanoptera, Thripidae) from the Afro-Tropical Region. *Zootaxa* **2010**, *2423*, 1–24. [CrossRef]
24. IBM Corp. *IBM SPSS Statistics for Windows, Version 25.0*; IBM Corp: Armonk, NY, USA, 2017.
25. Goldarazena, A.; Mound, L.A. Introducción a La Fauna de Los Tisanópteros (Cl. Insecta; O. Thysanoptera) de Navarra-Nafarroa, Sus Plantas Hospedadoras y Su Distribución. I Suborden Terebrantia. *Estudi. Mus. Cienc. Nat. Alava* **1997**, *12*, 167–202.
26. Priesner, H. *Ordnung Thysanoptera. Bestimmungsbüchenzur Bodenfauna Europas 2*; Akademie Verlag: Berlin, Germany, 1964.
27. Mound, L.A.; Morison, G.D.; Pitkin, B.R.; Palmer, J.M. *Thysanoptera. Handbooks for the Identification of British Insects*; Royal Entomological Society of London: London, UK, 1976; Volume 1.
28. Titschack, E. Zwei Neue Thysanopteren Aus Südeuropa. *Verh. Ver. Naturwiss. Heimatforsch.* **1958**, *33*, 4–15.

29. Lowry, V.K.; Smith Jr, J.W.; Mitchell, F.L. Life-Fertility Tables for *Frankliniella fusca* (Hinds) and *F. occidentals* (Pergande) (Thysanoptera: Thripidae) on Peanut. *Ann. Entomol. Soc. Am.* **1992**, *85*, 744–754. [CrossRef]
30. Bournier, A. *Les Thrips: Biologie, Importance Agronomique*; Inra: Paris, France, 1983.
31. Loomans, A.J.M.; Van Lenteren, J.C. *Biological Control of Thrips Pests: A Review on Thrips Parasitoids*; Wageningen Agricultural University: Wageningen, The Netherlands, 1995.
32. Etebari, K.; Matindoost, L.; Singh, R.N. Decision Tools for Mulberry Thrips Pseudo Dendrothrips Mori (Niwa, 1908) Management in Sericultural Regions; an Overview. *Insect Sci.* **2004**, *11*, 243–255. [CrossRef]
33. Thiéry, D.; Derridj, S.; Calatayud, P.A.; Maher, N.; Marion-Poll, F. L'insecte Au Contact Des Plantes. *Interacti. Insectes-Plantes, Partie* **2013**, *4*, 347–368.
34. Voorrips, R.E.; Steenhuis-Broers, G.; Tiemens-Hulscher, M.; van Bueren, E.T.L. Plant Traits Associated with Resistance to Thrips Tabaci in Cabbage (*Brassica oleracea* Var *capitata*). *Euphytica* **2008**, *163*, 409–415. [CrossRef]

Disclaimer/Publisher's Note: The statements, opinions and data contained in all publications are solely those of the individual author(s) and contributor(s) and not of MDPI and/or the editor(s). MDPI and/or the editor(s) disclaim responsibility for any injury to people or property resulting from any ideas, methods, instructions or products referred to in the content.

Article

Assessing the Impact of Variety, Irrigation, and Plant Distance on Predatory and Phytophagous Insects in Chili

András Lajos Juhász [1], Márk Szalai [2] and Ágnes Szénási [1,*]

[1] Department of Integrated Plant Protection, Plant Protection Institute, Hungarian University of Agriculture and Life Sciences, 2100 Gödöllő, Hungary
[2] Vak Bottyán str. 20, 2100 Gödöllő, Hungary
* Correspondence: szenasi.agnes@uni-mate.hu

Abstract: Chilies are plants that are becoming increasingly popular all over the world, including in Hungary. Since little is known about the abundance and seasonal dynamic of insect pests and their natural enemies associated with chilies under Hungarian climatic conditions, the aim of the study was to monitor these organisms on different varieties under different growing conditions to provide data for improving IPM for chilies. Chili varieties "Yellow Scotch Bonnet" (YSB) and "Trinidad Scorpion Butch T" (TSBT) were planted with three replicates. Two different plant-to-plant distances (30 vs. 40 and 40 vs. 60 cm in YSB, TSBT, respectively) and two different irrigation frequencies (daily, 40 min; every second day, 20 min) were used. Fifty flowers/plot/date were collected. In 2019, *Orius* (Hemiptera: Anthocoridae) larvae, and in 2021, phytophagous thrips larvae were dominant in all the treatments. Significantly more *Orius* adults and larvae were found in the YSB than in the TSBT variety and the number of *Aeolothrips* and phytophagous thrips (Thysanoptera: Thripidae) adults was significantly higher under less irrigation in 2019. The plant spacing did not affect the abundance of predators or herbivores. Upon comparing the two years, no effect of the treatments on the studied insect taxa was observed.

Keywords: *Capsicum chinense*; *Orius* sp.; *Aeolothrips* sp.; Thysanoptera; insect pest; natural enemies; irrigation frequency; plant spacing

1. Introduction

Capsicum species belonging to the Solanaceae family are among the most popular and widely consumed spice plants [1] and they have been cultivated for about 6000 years [2]. The worldwide production and harvested area of green peppers and chilies have increased in the last decade [3]. Possibly due to the capsaicinoid produced only by *Capsicum* species, peppers were used as a medicinal plant in ancient times [4]. Capsaicinoid extracts have medicinal properties, which can help to stop the pain of arthritis (rheumatoid arthritis and osteoarthritis) and arterial diseases [5]. Due to their pungency deriving from their capsaicinoid contents, chilies are very popular in the cuisines of America, Asia, Africa and Europe [6]. In addition, the green fruits of different pepper genotypes are rich sources of vitamins A, C and E [7]. Moreover, the first extraction of vitamin C was first successfully conducted in 1928 by Hungarian biochemist Albert Szent-Györgyi from sweet peppers, and this achievement led to him being awarded the Nobel Prize in physiology and medicine in 1937 [7].

Agrotechnological factors such as balanced irrigation and proper plant spacing help in the prevention of exposure to harmful organisms, an element of the first principle of integrated pest management (Directive 2009/128/EC) [8]. Due to their sensitivity to water stress [9], for optimal yield and quality, *Capsicum* species need an adequate water supply [10]. Drip irrigation systems are able to save water [11,12] and optimize the water supply, improving yield and delivering water directly to the roots [13]. Among agrotechnical methods, the manipulation of plant spacing is one of the most favored and promising ways of reducing pests [14]. Plant spacing and row spacing have important impacts on the plant growth, width and yield of chili pepper [15]. Above a certain plant density, nutrients, sunlight and water become less available because of the competition between plants [16]. Closer plant spacing generates higher relative humidity, which creates advantageous conditions for pathogens [16]. In the lower parts of plants, light leakage and aeration are heavily reduced by higher plant spacing and this also obstructs pollination [16] On the other hand, significantly more flowers and fruits were observed with an increase in plant density [15]. Therefore, a good compromise is necessary to determine the right plant spacing.

Although the rate of pestivorous species is only around 1% of the 6377 described known species in the order Thysanoptera [17,18], thrips are important insect pests in pepper, damaging different plant parts such as the leaves, flowers and fruits [19], as well as transmitting plant viruses such as Groundnut ringspot virus (GRSV), Tomato chlorotic spot virus (TCSV), Tomato spotted wilt virus (TSWV) and Watermelon silver mottle virus (WSMV) [20,21]. The following Thysanoptera species are reported as pests of *Capsicum* spp.: in Europe, *Franklinella occidentalis* (Pergande) and *Thrips tabaci* Lindeman; in North America, *Frankliniella bispinosa* (Morgan), *Frankliniella fusca* (Hinds), *F. occidentalis*, *Frankliniella tritici* (Fitch), and *T. tabaci*; in South Africa and South Asia, *Frankliniella schultzei* (Trybom); in Japan, Brazil and the Caribbean region, *Thrips palmi* Karny; in Java, Indonesia, *Thrips parvispinus* (Karny); and, furthermore, in India, *Scirtothrips dorsalis* Hood [19,22–25]. The identified thrips vectors of the abovementioned tospoviruses infecting *Capsicum* species are *F. bispinosa* (TSWV), *F. fusca* (TSWV), *Frankliniella intonsa* (Trybom) (TCSV and TSWV), *F. occidentalis* (GRSV, TCSV, and TSWV), *F. schultzei* (GRSV, TCSV, and TSWV), *T. palmi* (WSMV), *Thrips setosus* (Moulton) (TSWV) and *T. tabaci* (TSWV) [21]. Among Thysanoptera, beneficial species can also be found. Species belonging to the *Aeolothrips* genus are mainly flower-dwelling and predators [18]. The Hungarian Thysanoptera checklist consists of eight *Aeolothrips* species [26] and among them, *Aeolothrips intermedius* Bagnall is widespread in Europe [27] and frequent in Hungary as well [28], occurring in different biotopes, mainly on dicotyledonous herbaceous plants [29]. Pollen can serve as an important alternative food for *A. intermedius* [30,31].

Orius species (Heteroptera: Anthocoridae) are generalist predators, but they prefer Thysanoptera from the family Thripidae [32]. They are abundant all over the world, are able to invade field crops rapidly, and, furthermore, can effectively control the population of thrips species under field conditions [33–36] since they can prey on up to 10 thrips individuals/day [37]. *Orius niger* W. and *Orius minutus* L. are the most frequent species from this genus in Hungary) [38].

Boateng et al. (2017) [16] observed the hot pepper plant morphology and yield with various plant spacing (70 × 30, 70 × 40, and 70 × 50 cm) in Ghana. Setiawati et al. (2022) [15] studied the effects of different plant densities (20,000, 30,000 or 40,000 plants ha^{-1}) on disease and pest susceptibility in chilies in Indonesia. Among the arthropod pests, only thrips damage was reported, which was reduced with a decrease in plant density. Das et al. (2021) [14] observed an effect of plant spacing (50 × 60, 50 × 50, 50 × 40, and 50 × 30 cm) on the incidence of chili leaf curl virus in hot peppers in Bangladesh, but not on arthropod pests. O'Keefe and Palada (2002) [39] studied only the row spacing (41, 46 and 61 cm) in U.S. Virgin Islands, but no effect of the plant-to-plant distance on the yield and growth of chilies was observed, and they performed no arthropod observations. Karungi et al. (2013) [40] assessed the effect of plant (80 × 80 vs. 60 × 50 cm) density on aphids, thrips, whiteflies and viruses in the Scotch Bonnet hot chili variety in Uganda. Closer plant spacing decreased the aphid and whitefly abundances but not the abundance of thrips. There are no data available on the effects of irrigation on the insect density and population growth in chilies. Since little is known about the abundance and seasonal dynamics of insect pests and their natural enemies associated with chilies under European conditions, the aim of this study was to monitor these organisms on different varieties under different growing conditions to provide data for improving the IPM of chilies in Europe.

2. Materials and Methods

2.1. Experimental Design

Study was carried out at the experimental farm of the Institute of Horticultural Science, Hungarian University of Agriculture and Life Sciences, Gödöllő, Hungary (47.58 N, 19.37 E). In total, twenty-four plots were established on sandy clay loam soil. The plot size was determined according to the number of available seedlings. The same adjustments were carried out for both years. However, in 2021, due to the low plant germination rate and the weather circumstances in May (continuous cloudy weather and rainfall), the sizes of the plots and of the experimental area were smaller than those in 2019.

Two chili (*Capsicum chinense* Jacq.) varieties (Trinidad Scorpion Butch T, Yellow Scotch Bonnet) were planted in plots of 12.96 m^2 (3.6 × 3.6 m) in the year 2019 and 2.16 m^2 (1.8 m × 1.2 m) in the year 2021. Eight different combinations of settings were used. Each variety was planted in two different plant spacings according to their usual plant height values, and two different irrigation frequencies/variety by drip irrigation was used. The number of replicates was three. Since 61 cm was found to be the optimal row spacing in the study of O'Keefe and Palada (2002) [39], we decided to use 60 cm row spacing for both varieties in our experiment.

The coding of the treatments was as follows:

V1: TSBT
V2: YSB
V1PS1: Plant spacing of 60 cm
V1PS2: Plant spacing of 40 cm
V2PS1: Plant spacing of 40 cm
V2PS2: Plant spacing of 30 cm
I1: Irrigation daily, 40 min (7.33 L/m/day)
I2: Irrigation every second day, 20 min (3.66 L/m/two days)

Seeds of the YSB variety were obtained from Caribbean Garden Seed, USA, and those of the TSBT variety, from Samenchilishop, Germany. Seeds were sowed in seedling trays on 12 March in 2019 and 7 April in 2021 and were placed, first, in a germination chamber and, later, in a nursery. Seed dressing using spores of the *Pythium oligandrum* mycoparasite fungus was performed each year.

Seedlings were planted on 24 May 2019 and 25 June 2021. Plants were harvested on 11–12 September in 2019 and 11 October in 2021. After planting, no pesticide applications were used during the growing seasons.

2.2. Characteristics of the Studied Varieties

TSBT is one of the six hottest chilies of the world [4], its pungency is between 800,000 and 1,460,000 on the Scoville-scale [41]. The landrace Trinidad Scorpion originates from Trinidad and Tobago. 'Butch T' means that it is grown by Butch Taylor in Australia [42]. The plant height ranges between 100 and 200 cm. It is a fast-maturing variety since only 90 to 120 days are needed for its ripening [43], depending on environmental conditions. YSB is one of the most widespread chili varieties in the Caribbean region and is grown extensively in Jamaica. Its SHU (Scoville Heat Unit) value ranges from 200,000 to 350,000; a slow-growing, bushy variety, with a maximum height of 50 cm. For its ripening, 160 days are needed [44].

2.3. Meteorological Data

Weather data (daily minimum, maximum and average temperature and precipitation) were obtained from Meteoblue AG (Basel, Switzerland) for both vegetation periods (Figure 1).

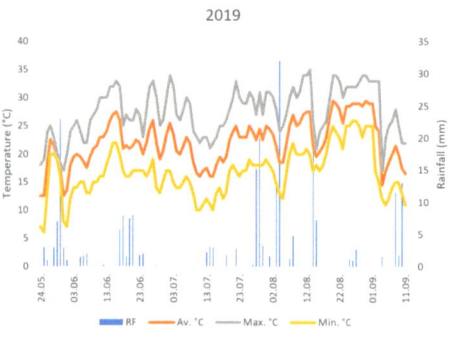

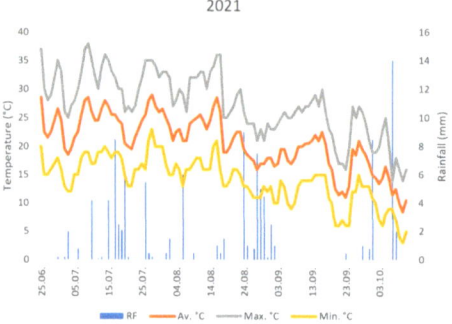

Figure 1. Daily maximum (Max.), minimum (Min.), and average (Av.) air temperature and rainfall (RF) data for the growing seasons 2019 and 2021.

2.4. Flower Sample Collection

Observation begun two weeks later than start of flowering. In both years, ten plants per plot were randomly selected and five flowers per plant were collected and stored in a plastic tube containing 60% propanol. Thysanoptera (phytophagous and predatory species) and *Orius* adults and larvae were counted using a stereomicroscope.

2.5. Statistical Analysis

Data were aggregated and analyzed using R [45] with Rcmdr package [46]. Generalized linear models were used to investigate individually the effect of variety/irrigation frequency/plant spacing as single explanatory variables on the abundance of predatory bugs (*Orius* spp.) and phytophagous and predatory Thysanoptera individuals. For all the hypothesis tests, a threshold of alpha = 5% was used to control type I error. However, there was no specific control factor for beta / type II error. However, we used the tests with the highest power possible (considering the assumptions), and this way we could minimize type II error. Moreover, we fitted models individually for variety/irrigation frequency/plant spacing as single explanatory variables on abundances, and these variables had two levels. Therefore, no *p*-value adjustment was performed for multiple comparisons. In the case of a low number of individuals, Poisson models were fitted; otherwise, Gaussian models. Assumptions of residual normality and homoscedasticity as well as potential influential data points were checked with basic model diagnostic plots [47].

3. Results

3.1. Abundance of Predatory and Phytophagous Insects by Treatments

Altogether, 117 *Aeolothrips* adults, 160 *Aelothrips* larvae, 316 *Orius* adults, 1451 *Orius* larvae, 805 phytophagous thrips adults and 1287 phytophagous thrips larvae were collected in chili flowers during the two years. In 2019, *Orius* larvae and phytophagous thrips adults were the most numerous taxa, while in 2021, phytophagous thrips larvae and *Orius* larvae were the most abundant taxa (Table 1 and Table S1). The individual numbers of each taxon were similar for the different treatments. In the first year, the most specimens were found in the plots with less irrigation, in the second vegetation period, in the plots with increased plant spacing (Table S1).

Table 1. Total insect numbers in chili flowers during growing seasons.

Year	*Aeolothrips* Adults	*Aeolothrips* Larvae	*Orius* Adults	*Orius* Larvae	Phytophagous Thrips Adults	Phytophagous Thrips Larvae
2019	103	116	146	877	512	213
2021	14	44	170	574	293	1030

In the year 2019, significantly more *Orius* adults ($p < 0.001$) and larvae ($p = 0.0135$) were found in the flowers of the YSB variety, but the numbers of these insects were similar under the different irrigation treatments and plant spacing (Figure 2A). The abundances of phytophagous thrips larvae did not differ between the treatments (Figure 2B). As for the number of all the *Orius* individuals (larvae and adults together), a significant difference between the two varieties was detected ($p = 0.00119$, Figure 2C). The numbers of all phytophagous thrips (larvae and adults together), phytophagous thrips and *Aeolothrips* adults were significantly higher under less irrigation ($p = 0.00916$, $p = 0.0098$, $p < 0.001$, respectively), while the abundances of *Aeolothrips* larvae did not differ between the treatments (Figure 2B–D, also see Table 2).

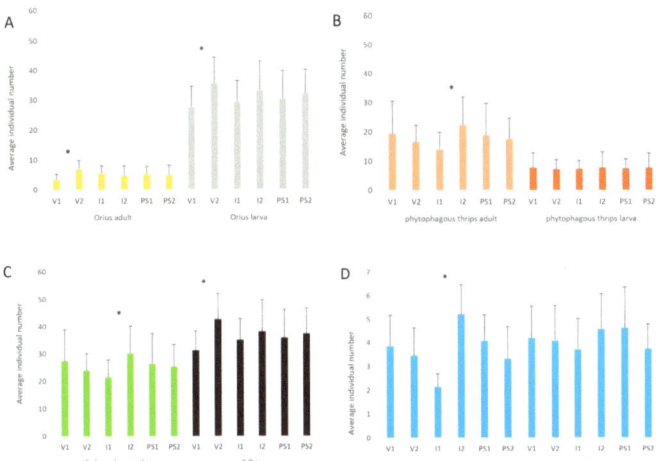

Figure 2. Average individual numbers of predatory and phytophagous insects in chili flowers under treatments in 2019: (**A**) *Orius* spp. (**B**) Phytophagous thrips. (**C**) All *Orius* and phytophagous thrips (adults and larvae). (**D**) *Aeolothrips* spp. (V1) Trinidad Scorpion Butch T; (V2) Yellow Scotch Bonnet; (I1) Irrigation daily, 40 min; (I2) Irrigation every second day, 20 min; (PS1) Increased plant spacing; (PS2) Decreased plant spacing. * Denotes significant differences between the treatment levels ($p < 0.001$). The error bars represent the standard errors.

In the year 2021, the numbers of phytophagous thrips larvae were significantly higher ($p = 0.0266$) under more irrigation, while the abundances of all the other taxa did not differ between the treatments (Figure 3A–D; also, see Table 3). Upon pooling the two years' data, no effects of the treatments on insect taxa were observed.

Table 2. Effects of variety, irrigation level and plant spacing on the insect abundance of chili flowers in 2019. Residual degrees of freedom were 138 in all tests.

Insect Taxa	Variety			Irrigation			Plant Spacing		
	Effect Estimate	Test Statistics	p Value	Effect Estimate	Test Statistics	p Value	Effect Estimate	Test Statistics	p Value
Phytophagous thrips adults	2.8307	F = 0.8876	0.4160	−8.2857	F = 8.0624	0.0099	1.3333	F = 0.4139	0.7030
Phytophagous thrips larvae	0.5589	χ^2 = 0.2866	0.8660	−0.3571	χ^2 = 0.1174	0.3950	−0.2717	χ^2 = 0.0677	0.4005
All phytophagous thrips	3.3897	F = 1.1328	0.2189	−8.6428	F = 7.8629	0.0091	1.0615	F = 0.2804	0.6013
Orius adults	−3.4769	χ^2 = 16.190	<0.001	0.7142	χ^2 = 0.6855	0.4080	0.3179	χ^2 = 0.1348	0.7130
Orius larvae	−7.8717	F = 0.0149	0.0135	−3.78572	F = 0.2915	0.4460	−1.7487	F = 0.2430	0.7550
All *Orius*	−11.3487	F = 12.5463	0.0011	−3.0714	F = 0.5030	0.4460	−1.4307	F = 0.1434	0.7550
Aeolothrips adults	0.4051	χ^2 = 0.3117	0.5810	−3.0714	χ^2 = 18.513	<0.001	0.7435	χ^2 = 1.0435	0.3080
Aeolothrips larvae	0.1230	χ^2 = 0.0255	0.8100	−0.8571	χ^2 = 1.2436	0.2260	0.8820	χ^2 = 1.3037	0.2210
All *Aeolothrips*	0.5282	χ^2 = 0.2489	0.4441	−3.9285	χ^2 = 13.962	<0.001	1.6256	χ^2 = 2.3457	0.0714

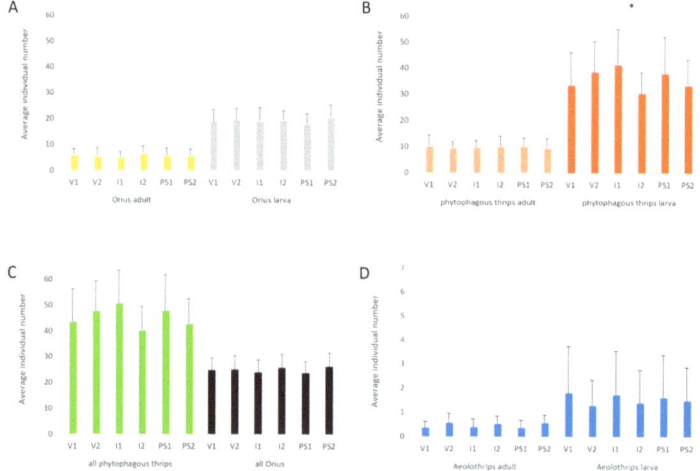

Figure 3. Average individual numbers of predatory and phytophagous insects in chili flowers under treatments in 2021: (**A**) *Orius* spp. (**B**) Phytophagous thrips. (**C**) All *Orius* and phytophagous thrips (adults and larvae). (**D**) *Aeolothrips* spp. (V1)Trinidad Scorpion Butch T; (V2) Yellow Scotch Bonnet; 230 (I1) Irrigation daily, 40 min; (I2) Irrigation every second day, 20 min; (PS1) Increased plant spacing; (PS2) Decreased plant spacing. * Denotes significant differences between the treatment levels (p < 0.001). The error bars represent the standard errors.

3.2. Seasonal Dynamics

The average numbers of individuals for all the taxa except for *Orius* larvae; for some of the dates, *Aeolothrips* larvae decreased to August; and the insect abundance was always below 10 specimens in all the treatments. For almost all the dates and treatments, more phytophagous thrips adults were found in the flowers than larvae (Figure 4). The seasonal dynamics of certain insect taxa were similar in the flowers of both varieties (Figure 4A,B); however, on almost all the dates, more thrips adults were collected in the less-irrigated plots than under increased irrigation (Figure 4C,D). Except for on a single date, the number of phytophagous thrips larvae and *Orius* larvae was always slightly higher under decreased plant spacing than under a higher plant density (Figure 4E,F).

Table 3. Effects of variety, irrigation level and plant spacing on the insect abundance in chili flowers in 2021. Residual degrees of freedom were 178 in all tests.

Insect Taxa	Variety			Irrigation			Plant Spacing		
	Effect Estimate	Test Statistics	p Value	Effect Estimate	Test Statistics	p Value	Effect Estimate	Test Statistics	p Value
Phytophagous thrips adults	0.7678	$\chi^2 = 0.4517$	0.3620	−0.3333	$\chi^2 = 0.0853$	0.7850	0.6339	$\chi^2 = 0.3078$	0.4220
Phytophagous thrips larvae	−5.0625	F = 1.5248	0.4383	10.8000	F = 6.9704	0.0266	4.5803	F = 1.0643	0.2829
All phytophagous thrips	−4.2946	F = 1.1472	0.6198	10.4666	F = 6.8442	0.0561	5.2142	F = 1.5056	0.2637
Orius adults	0.4464	$\chi^2 = 0.2631$	0.5150	−1.6000	$\chi^2 = 3.3996$	0.0608	0.0446	$\chi^2 = 0.0026$	0.8250
Orius larvae	−0.6875	F = 0.1674	0.7960	−0.2666	F = 0.0253	0.8530	−2.5625	F = 2.4276	0.1770
All *Orius*	−0.2411	F = 0.0170	0.9330	−1.8666	F = 1.0233	0.3000	−2.5178	F = 1.8886	0.2800
Aeolothrips adults	−0.1964	$\chi^2 = 0.6161$	0.4240	−0.1333	$\chi^2 = 0.2867$	0.5820	−0.1964	$\chi^2 = 0.6161$	0.4240
Aeolothrips larvae	0.5267	$\chi^2 = 1.339$	0.2098	0.3333	$\chi^2 = 0.5329$	0.4552	0.1250	$\chi^2 = 0.0746$	0.6571
All *Aeolothrips*	0.3303	$\chi^2 = 0.4027$	0.4707	0.2000	$\chi^2 = 0.1476$	0.6950	−0.0714	$\chi^2 = 0.0187$	0.9959

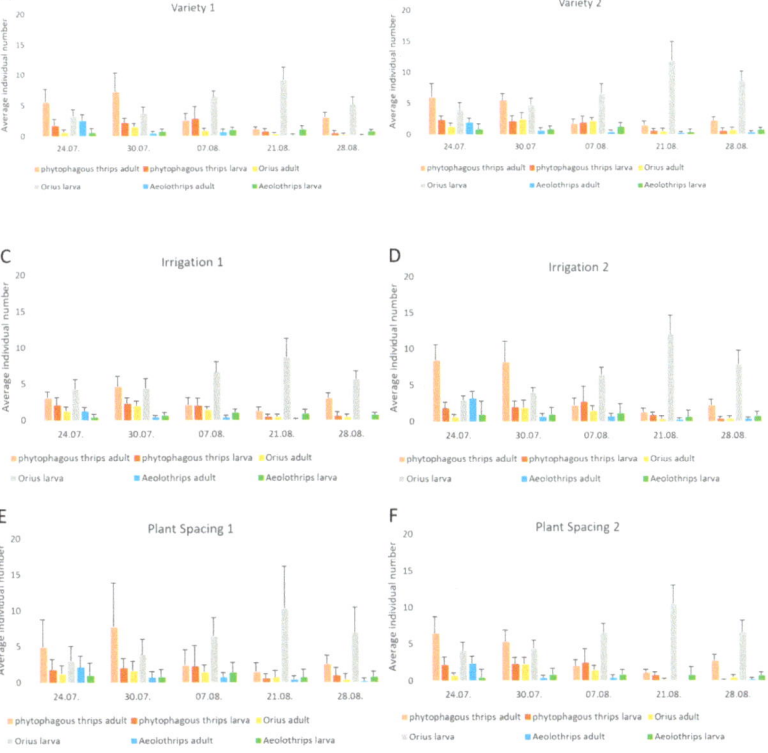

Figure 4. Seasonal dynamics of predatory and phytophagous insects in chili pepper flowers in 2019. (**A**) Trinidad Scorpion Butch T. (**B**) Yellow Scotch Bonnet. (**C**) Irrigation daily, 40 min. (**D**) Irrigation every second day, 20 min. (**E**) Increased plant spacing. (**F**) Decreased plant spacing. The error bars represent the standard errors.

Phytophagous thrips larvae were dominant, except on 13 September, and *Aeolothrips* adults were the less-numerous insects in the vegetation period in all the treatments. Besides phytophagous thrips larvae, only *Orius* larvae could reach five individuals on some dates, and the number of other taxa was always below this value. On all the dates, more *Orius* larvae were present in the flowers than adults (Figure 5). The seasonal dynamics of certain insect groups were similar in the flowers of both varieties (Figure 5A,B). On almost all the dates, more *Orius* adults and larvae but fewer phytophagous thrips were found under less irrigation than in the more frequently irrigated plots (Figure 5C,D). Except for on two dates, the larval number of phytophagous thrips and *Orius* spp. was always slightly higher under decreased plant spacing than under an increased plant density (Figure 5E,F).

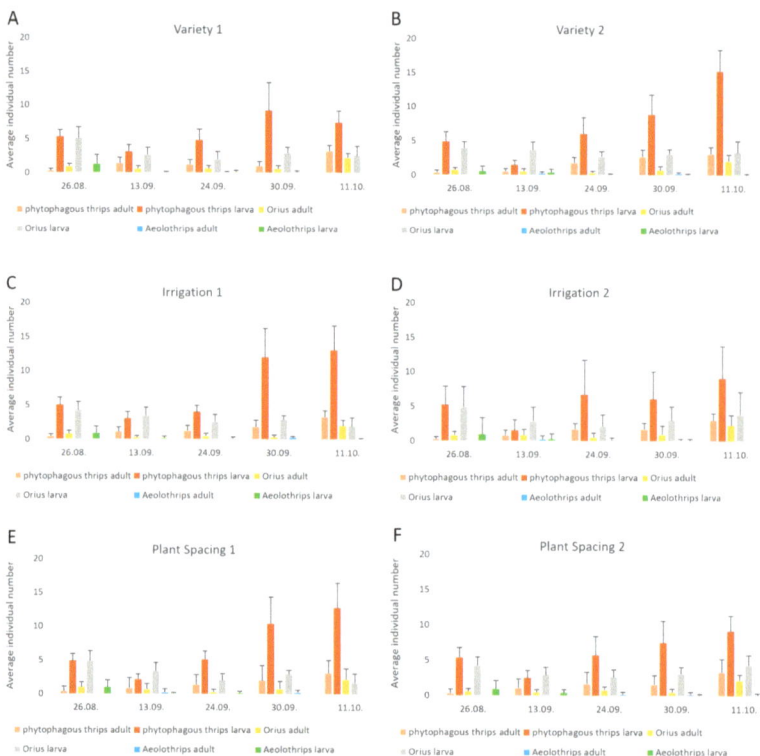

Figure 5. Seasonal dynamics of predatory and phytophagous insects in chili pepper flowers in 2021. (**A**) Trinidad Scorpion Butch T. (**B**) Yellow Scotch Bonnet. (**C**) Irrigation daily, 40 min. (**D**) Irrigation every second day, 20 min. (**E**) Increased plant spacing. (**F**) Decreased plant spacing. The error bars represent the standard errors.

4. Discussion

This is the first time the insect communities of chili flowers in the European region have been studied. The literature data on the arthropods found in chili flowers are relatively scarce; moreover, simultaneous data collection for different irrigation frequencies and plant spacing is also limited.

Temperature and rainfall are the most important weather factors affecting thrips populations under field conditions. Warm, sunny and dry weather in the summer are favorable for the survival and population growth of most thrips species in temperate zones [48]. Probably due to this circumstance, the number of Thysanoptera adults and all phytophagous thrips was significantly higher under decreased irrigation than under increased irrigation, but only in the first year. Furthermore, the largest abundance of these taxa was detected in the less-irrigated plots. On the contrary, the phytophagous thrips abundance was similar at higher and lower irrigation levels in 2021. No effect of the irrigation frequency on *Orius* spp was observed.

Effects of the studied varieties were detected only on predatory bugs. The abundance of *Orius* spp. (larvae, adults and together) was significantly higher, whereas the thrips abundance (regardless of the species or developmental stages) was significantly lower on the YSB variety in 2019. The number of *Orius* and Thysanoptera individuals on the same variety was similar in the second study year.

No effect of the varieties on phytophagous thrips was found, since the adult and larval numbers of these insects on the two varieties were similar in 2019 and in 2021. On the other hand, different thrips resistance levels have been detected in different *Capsicum* species (*C. annuum, C. chinense, C. baccatum* and *C. frutescens*) and varieties [19,49,50]. Moreover, there are also chili pepper varieties that are resistant to other animal pests such as the oriental fruit fly (*Bactrocera dorsalis*) [51,52] and the southern root-knot nematode (*Meloidogyne incognita*) [53].

Two different plant densities (60 × 50 and 80 × 80 cm) were performed with a 4 × 8 m plot size for the Scotch Bonnet (*Capsicum chinense*) variety, and the plant spacing had no significant effect, but the vegetation period and sampling date had a significant effect on phytophagous thrips abundance [40]. Setiawati et al. (2022) [15] found that plant density could effectively influence chili pepper quality and thrips damage. Three different plant densities (20,000, 30,000 and 40,000 plants ha^{-1}) with 13 × 14 m plot size were used. Less thrips damage was detected under the lowest plant density, followed by densities of 30,000 and 40,000 plants ha^{-1}. On the contrary, our results indicated that the insect abundances were similar under different plant spacing both years.

The flower, as a habitat, may ensure protection from rainfall, unfavorable temperatures, solar radiation and (due to the higher relative humidity) desiccation for the thrips species [54]. However, the number of suitable flowers within an area might influence thrips population density, and among others, due to this circumstance, Thysanoptera species have great mobility within and between habitats [48]. In our studies, beside phytophagous thrips individuals, *Orius* and *Aeolothrips* species were found in chili flowers, which are among the primary predators of Thysanoptera [48]. Similarly, in field melons primarily infested with thrips, the main predators in the flowers were *Orius* spp. (67.4%) and *Aeolothrips* spp. (32.6%) [55].

Similarly to the results of Sabelis and Van Rijn (1997), Funderburk et al. (2000) and Ramachandran et al. (2001) [33–35], we also found that *Orius* predatory bugs were abundant and could effectively control the individual number of Thysanoptera species under field conditions in the first year. It was observed that predators were responding to the number of *Thrips tabaci* individuals in onions [56], and, in parallel to this, our experience was that the *Orius* larval number was increased in the vegetation period of the year 2019, whereas the phytophagous thrips abundance was decreased. However, an opposite trend was detected in 2021, and the *Orius* population was not be able to follow the prey abundance. On the other hand, this type of response of *Aeolothrips* species could not be observed, probably because of their low abundance. The number of this predatory organism declined progressively in 2019, and few adults were found in the next season. Although *Orius* species prefer Thysanoptera from the Thripidae family [32], *Orius albidipennis* was able to prey on *Aeolothrips fasciatus* from the Aeolothripidae family and reduce its number in the field [57]. Furthermore, intraguild predation was discovered between *Orius niger* and *Aeolothrips intermedius* [58]. Therefore, *Orius* species may also regulate predatory thrips populations in chili flowers. However, in our observations, *Aeolothrips* individuals were much less abundant than *Orius* specimens, especially in the second year.

Since plant genetic diversity is an important element of integrated plant protection for chilies [59], the investigation of more *Capsicum* species and varieties in future research will be worthwhile.

5. Conclusions

Based on our trials with chili cultivation practices, we conclude that a higher drip irrigation frequency and water amount negatively impacted the number of adults of phytophagous and predatory Thysanoptera in the first year of the study. These results might be considered in chili cultivation, depending on the environmental conditions, e.g., the season and soil type with special regard to the climate. The chili varieties in our study resulted in an impact on the density of the predatory *Orius* species but not that of other insects species; however, further tests on other varieties should also be assessed. We found that the plant spacing of the chilies did not affect the abundance of predators or herbivores. According to our observations, *Orius* predatory bugs are key species in chili cultivation; therefore, the protection of these predators as non-target organisms in the IPM of chilies is an important aspect to be considered. Our results provide important support for the development of IPM for chilies.

Supplementary Materials: The following supporting information can be downloaded at: https://www.mdpi.com/article/10.3390/horticulturae8080741/s1, Table S1: Total individual numbers of insects in chili flowers under treatments.

Author Contributions: Conceptualization, Á.S. and A.L.J.; methodology, Á.S. and A.L.J.; investigation, A.L.J.; data curation, M.S. and A.L.J.; writing—original draft preparation, Á.S., M.S. and A.L.J.; writing—review and editing, Á.S., M.S. and A.L.J.; visualization, A.L.J., M.S. and Á.S.; supervision, Á.S. All authors have read and agreed to the published version of the manuscript.

Funding: This research received no external funding.

Institutional Review Board Statement: Not applicable.

Informed Consent Statement: Not applicable.

Data Availability Statement: The data are available on request due to institutional restrictions and privacy.

Acknowledgments: The authors would like to thank everyone who contributed to this article and the field and laboratory work.

Conflicts of Interest: The authors declare no conflict of interest.

References

1. Garcés-Claver, A.; Arnedo-Andrés, M.S.; Abadía, J.; Gil-Ortega, R.; Álvarez-Fernandez, A. Determination of capsaicin and dihydrocapsaicin in *Capsicum* fruits by liquid chromatography–electrospray/time-of-flight mass spectrometry. *J. Agric. Food Chem.* **2006**, *54*, 9303–9311. [CrossRef]
2. Perry, L.; Dickau, R.; Zarrillo, S.; Holst, I.; Pearsall, D.M.; Piperno, D.R.; Berman, M.J.; Cooke, R.G.; Rademaker, K.; Ranere, A.J.; et al. Starch fossils and the domestication and dispersal of chili peppers (*Capsicum* spp. L.) in the Americas. *Science* **2007**, *315*, 986–988. [CrossRef]
3. FAOSTAT. 2022. Available online: https://www.fao.org/faostat/en/#data/QCL/visualize (accessed on 1 July 2022).
4. Omolo, M.A.; Wong, Z.Z.; Mergen, A.K.; Hastings, J.C.; Le, N.C.; Reiland, H.A.; Case, K.A.; Baumler, D.J. Antimicrobial properties of chili peppers. *J. Infect. Dis. Ther.* **2014**, *2*, 145. [CrossRef]
5. Cordell, G.A.; Araujo, O.E. Capsaicin: Identification, nomenclature, and pharmacotherapy. *Ann. Pharmacother.* **1993**, *27*, 330–336. [CrossRef] [PubMed]
6. Guzmán, I.; Bosland, P.W. Sensory properties of chile pepper heat—And its importance to food quality and cultural preference. *Appetite* **2017**, *117*, 186–190. [CrossRef] [PubMed]
7. Kumar, S.; Kumar, R.; Singh, J. Cayenne/American Pepper. In *Handbook of Herbs and Spices*, 1st ed.; Peter, K.V., Ed.; Woodhead Publishing: Cambridge, UK, 2006; Volume 3, pp. 299–312.
8. Barzman, M.; Bàrberi, P.; Birch, A.N.; Boonekamp, P.; Dachbrodt-Saaydeh, S.; Graf, B.; Hommel, B.; Jensen, J.; Kiss, J.; Kudsk, P.; et al. Eight principles of integrated pest management. *Agron. Sustain. Dev.* **2015**, *35*, 1199–1215. [CrossRef]
9. González-Dugo, V.; Orgaz, F.; Fereres, E. Responses of pepper to deficit irrigation for paprika production. *Sci. Hort.* **2007**, *114*, 77–82. [CrossRef]
10. Delfine, S.; Tognetti, R.; Loreto, F.; Alvino, A. Physiological and growth responses to water stress in field-grown bell pepper (*Capsicum annuum* L.). *J. Hort. Sci. Biotechnol.* **2002**, *77*, 697–704. [CrossRef]
11. Singandhupe, R.B.; Rao, G.G.S.N.; Patil, N.G.; Brahmanand, P.S. Fertigation studies and irrigation scheduling in drip irrigation system in tomato crop (*Lycopersicon esculentum* L.). *Eur. J. Agron.* **2003**, *19*, 327–340. [CrossRef]
12. Maisiri, N.; Senzanje, A. On farm evaluation of the effect of low cost drip irrigation on water and crop productivity compared to conventional surface irrigation system. *Phys. Chem. Earth* **2005**, *30*, 783–791. [CrossRef]

13. Barkunan, S.R.; Bhanumathi, V.; Sethuram, J. Smart sensor for automatic drip irrigation system for paddy cultivation. *Comput. Electr. Eng.* **2019**, *73*, 180–193. [CrossRef]
14. Das, S.; Rahman, M.; Dash, P.K.; Kamal, M. Suppression of chili leaf curl virus (ChLCV) incidence in chili (*Capsicum annuum* L.) across Bangladesh via manipulated planting date and spacing. *J. Plant Dis. Prot.* **2021**, *128*, 535–548. [CrossRef]
15. Setiawati, W.; Muharam, A.; Hasyim, A.; Prabaningrum, L.; Moekasan, T.K.; Murtiningsih, R.; Lukman, L.; Mejaya, M.J. Growth, yield characters and pest and diseases severity of chili pepper under different plant density and pruning levels. *Appl. Ecol. Environ. Res.* **2022**, *20*, 543–553. [CrossRef]
16. Boateng, E.; Adjei, E.A.; Osei, M.K.; Offei, K.O.; Olympio, N.S. Response of plant spacing on the morphology and yield of five hot pepper lines. *Afr. J. Agric. Res.* **2021**, *7*, 1281–1287. [CrossRef]
17. Morse, J.G.; Hoddle, M.S. Invasion biology of thrips. *Ann. Rev. Entomol.* **2006**, *51*, 67–89. [CrossRef] [PubMed]
18. ThripsWiki. ThripsWiki-Providing Information on the World's Thrips. Available online: http://thrips.info/wiki/ (accessed on 29 June 2022).
19. Maharijaya, A.; Vosman, B.; Steenhuis-Broers, G.; Harpenas, A.; Purwito, A.; Visser, R.G.F.; Voorrips, R.E. Screening of pepper accessions for resistance against two thrips species (*Frankliniella occidentalis* and *Thrips parvispinus*). *Euphytica* **2011**, *177*, 401–410. [CrossRef]
20. Roggero, P.; Pennazio, S.; Masenga, V.; Tavella, L. Resistance to tospoviruses in pepper. In Proceedings of the 7th International Symposium on Thysanoptera, Thrips and Tospoviruses, Reggio Calabria, Italy, 2–7 July 2001; Marullo, R., Mound, L., Eds.; pp. 105–110.
21. Whitfield, A.E.; Ullman, D.E.; German, T.L. Tospovirus–thrips interactions. *Ann. Rev. Phytopathol.* **2005**, *43*, 459–489. [CrossRef]
22. Lewis, T.; Mound, L.A.; Nakahara, S.; Childers, C.C. Major crops infested by thrips with main symptoms and predominant injurious species. In *Thrips as Crop Pests*, 1st ed.; Lewis, T., Ed.; CAB Int.: Wallingford, UK, 1997; pp. 675–709.
23. Reitz, S.R.; Yearby, E.L.; Funderburk, J.E.; Stavisky, J.; Momol, M.T.; Olson, S.M. Integrated management tactics for *Frankliniella* Thrips (Thysanoptera: Thripidae) in field-grown pepper. *J. Econ. Entomol.* **2003**, *96*, 1201–1214. [CrossRef]
24. Cannon, R.J.C.; Matthews, L.; Collins, D.W. A review of the pest status and control options for *Thrips palmi*. *Crop Prot.* **2007**, *26*, 1089–1098. [CrossRef]
25. Sahu, P.S.; Kumar, A.; Khan, H.H.; Naz, H. Seasonal incidence of chilli thrips, *Scirtothrips dorsalis* (Hood) (Thripidae: Thysanoptera): A review. *J. Entomol. Zool. Stud.* **2018**, *6*, 1738–1740.
26. Jenser, G. A checklist of Thysanoptera of Hungary. *Folia Entomol. Hung.* **2011**, *72*, 31–46.
27. zur Strassen, R. *Die terebranten Thysanopteren Europas und des Mittelmeer-Gebietes. Die Tierwelt Deutschlands 74*, 1st ed.; Verlag Goecke & Evers: Keltern, Germany, 2003; Volume 74, p. 46.
28. Jenser, G. Tripszek—Thysanoptera. In *Magyarország Állatvilága, Fauna Hungariae*, 1st ed.; Academic Press: Budapest, Hungary, 1982; Volume 13, p. 19.
29. Czencz, K.; Jenser, G. Microhabitats of *Aeolothrips intermedius* BAGNALL 1934 (Thysanoptera: Aeolothripidae). *Cour. Forschungsinst. Senckenberg.* **1994**, *178*, 51–55.
30. Trdan, S.; Andjus, L.; Raspudić, E.; Kač, M. Distribution of *Aeolothrips intermedius* Bagnall (Thysanoptera: Aeolothripidae) and its potential prey Thysanoptera species on different cultivated host plants. *J. Pest. Sci.* **2005**, *78*, 217–226. [CrossRef]
31. Trdan, S.; Rifelj, M.; Valič, N. Population dynamics of banded thrips (*Aeolothrips intermedius* Bagnall, Thysanoptera, Aeolothripidae) and its potential prey Thysanoptera species on white clover. *Comm. App. Biol. Sci. Ghent Univ.* **2005**, *70*, 753–758.
32. Baez, I.; Reitz, S.R.; Funderburk, J.E.; Reitz, S.R.; Funderburk, J.E. Predation by *Orius insidiosus* (Heteroptera: Anthocoridae) onlife stages and species of *Frankliniella* flower thrips (Thysanoptera: Thripidae) in pepper flowers. *Environ. Entomol.* **2004**, *33*, 662–670. [CrossRef]
33. Sabelis, M.W.; Van Rijn, P.C.J. Predation by insects and mites. In *Thrips as Crop Pests*, 1st ed.; Lewis, T., Ed.; CAB International: Wallingford, UK, 1997; pp. 259–354.
34. Funderburk, J.; Stavisky, J.; Olson, S. Predation of *Frankliniella occidentalis* (Thysanoptera: Thripidae) in field peppers by *Orius insidiosus* (Hemiptera: Anthocoridae). *Environ. Entomol.* **2000**, *29*, 376–382. [CrossRef]
35. Ramachandran, S.; Funderburk, J.; Stavisky, J.; Olson, S. Population abundance and movement of *Frankliniella* species and *Orius insidiosus* in field pepper. *Agric. Forest Entomol.* **2001**, *3*, 1–10. [CrossRef]
36. Loomans, A. Parasitoids as Biological Control Agents of Thrips Pests. Ph.D. Thesis, Wageningen University, Wageningen, The Netherlands, 2003.
37. Blümel, S.; Fischer-Colbrie, P.; Höbaus, E. *Nützlinge: Helfer im zeitgemässen Pflanzenschutz*, 3rd ed.; Verlag Jugend & Volk Ges.m.b.H.: Wien, Austria, 1988; p. 111.
38. Rácz, V. Poloskák (Heteroptera) Szerepe Magyarországi Kukoricások Életközösségében. Master's Thesis, Hungarian Academy of Science, Budapest, Hungary, 1989.
39. O'Keefe, D.A.; Palada, M.C. In-row plant spacing affects growth and yield of four hot pepper cultivars. In Proceedings of the Caribbean Food Crops Society, 38th Annual Meeting, Trois-Ilets, Martinique, France, 30 June–5 July 2002; Merlini, X., Jean-Baptiste, I., Mbolidi-Baron, H., Eds.; AMADEPA: Martinique, France; pp. 162–168.
40. Karungi, J.; Obua, T.; Kyamanywa, S.; Mortensen, C.N.; Erbaugh, M. Seedling protection and field practices for management of insect vectors and viral diseases of hot pepper (*Capsicum chinense* Jacq.) in Uganda. *Int. J. Pest. Manag.* **2013**, *59*, 103–110. [CrossRef]

41. Werner, J. Capsaicinoids—Properties and mechanisms of pro-health action. In *Analytical Methods in the Determination of Bioactive Compounds and Elements in Food. Food Bioactive Ingredients*; Jeszka-Skowron, M., Zgoła-Grześkowiak, A., Grześkowiak, T., Ramakrishna, A., Eds.; Springer: Cham, Switzerland, 2021; pp. 193–225. [CrossRef]
42. Bosland, P.W.; Coon, D.; Reeves, G. 'Trinidad Moruga Scorpion' pepper is the world's hottest measured chile pepper at more than two million Scoville heat units. *Horttechnology* **2012**, *22*, 534–538. [CrossRef]
43. Trinidad Scorpion Butch, T. Available online: https://pepperhead.com/shop/trinidad-scorpion-butch-t/ (accessed on 25 July 2022).
44. Maguire, K. *Red Hot Chilli Grower*, 1st ed.; Mitchell Beazley: London, UK, 2015; p. 110.
45. R Core Team. *R: A Language and Environment for Statistical Computing*; R Foundation for Statistical Computing: Vienna, Austria, 2021; Available online: https://www.R-project.org/ (accessed on 20 January 2022).
46. Fox, J. *Using the R Commander: A Point-and-Click Interface for R*; Chapman & Hall/CRC: Boca Raton, FL, USA, 2017; pp. 1–233.
47. Faraway, J.J. *Linear Models with R*, 2nd ed.; Chapman & Hall/CRC: Boca Raton, FL, USA, 2014; pp. 1–270.
48. Lewis, T. *Thrips. Their Biology, Ecology and Economic Importance*; Academic Press: London, UK, 1973; pp. 187–201.
49. Maris, P.C.; Joosten, N.N.; Peters, D.; Goldbach, R.W. Thrips resistance in pepper and its consequences for the acquisition and inoculation of *Tomato spotted wilt virus* by the Western Flower Thrips. *Phytopathology* **2003**, *93*, 96–101. [CrossRef]
50. Van Haperen, P.; Voorrips, R.E.; van Loon, J.J.A.; Vosman, B. The effect of plant development on thrips resistance in *Capsicum*. *Arthropod Plant Interact.* **2019**, *13*, 11–18. [CrossRef]
51. Syamsudin, T.S.; Faizal, A.; Kirana, R. Dataset on antixenosis and antibiosis of chili fruit by fruit fly (*Bactrocera dorsalis*) infestation. *Data Brief* **2019**, *23*, 103758. [CrossRef] [PubMed]
52. Syamsudin, T.S.; Kirana, R.; Karjadi, A.K.; Faizal, A. Characteristics of chili (*Capsicum annuum* L.) that are resistant and susceptible to oriental fruit fly (*Bactrocera dorsalis* Hendel) infestation. *Horticulturae* **2022**, *8*, 314. [CrossRef]
53. Fery, R.L.; Thies, J.A. Evaluation of *Capsicum chinense* Jacq. Cultigens for resistance to the Southern Root-knot Nematode. *Hort. Sci.* **1997**, *32*, 923–926. [CrossRef]
54. Kirk, W.D.J. Distribution, Abundance and Population Dynamics. In *Thrips as Crop Pests*; Lewis, T., Ed.; CAB Int.: Wallingford, UK, 1997; pp. 217–257.
55. de Pedro, L.; López-Gallego, E.; Pérez-Marcos, M.; Ramírez-Soria, M.J.; Sanchez, J.A. Native natural enemies in Mediterranean melon fields can provide levels of pest control similar to conventional pest management with broad-spectrum pesticides. *Biol. Control* **2021**, *164*, 104778. [CrossRef]
56. Fok, E.J.; Petersen, J.D.; Nault, B.A. Relationships between insect predator populations and their prey, *Thrips tabaci*, in onion fields grown in large-scale and small-scale cropping systems. *BioControl* **2014**, *59*, 739–748. [CrossRef]
57. El-Serwiy, S.A.; Razoki, I.A.; Ragab, A.S. Population density of *Thrips tabaci* (Lind.) and the predators *Orius albidipennis* (Reut.) and *Aeolothrips fasciatus* (L.) on onion. *J. Agric. Water Resour. Res.* **1985**, *4*, 57–67.
58. Fathi, S.A.A.; Asghari, A.; Sedghi, M. 2008. Interaction of *Aeolothrips intermedius* and *Orius niger* in controlling *Thrips tabaci* on potato. *Int. J. Agri. Biol.* **2008**, *10*, 521–525.
59. Abdala-Roberts, L.; Berny-Mier y Terán, J.C.; Moreira, X.; Durán-Yáñez, A.; Tut-Pech, F. Effects of pepper (*Capsicum chinense*) genotypic diversity on insect herbivores. *Agric. Forest Entomol.* **2015**, *17*, 433–438. [CrossRef]

Article

Characterization of Volatile Compounds from Tea Plants (*Camellia sinensis* (L.) Kuntze) and the Effect of Identified Compounds on *Empoasca flavescens* Behavior

Fani Fauziah [1,2], Agus Dana Permana [3] and Ahmad Faizal [1,*]

1 Plant Science and Biotechnology Research Group, School of Life Sciences and Technology, Institut Teknologi Bandung, Jalan Ganeca 10, Bandung 40132, Indonesia; fani_fauziah@ymail.com
2 Indonesia Research Institute for Tea and Cinchona (IRITC) Gambung, Bandung 40972, Indonesia
3 Agrotechnology and Bioproduct Engineering Research Group, School of Life Sciences and Technology, Institut Teknologi Bandung, Jalan Ganesa 10, Bandung 40132, Indonesia; agus@sith.itb.ac.id
* Correspondence: afaizal@sith.itb.ac.id

Citation: Fauziah, F.; Permana, A.D.; Faizal, A. Characterization of Volatile Compounds from Tea Plants (*Camellia sinensis* (L.) Kuntze) and the Effect of Identified Compounds on *Empoasca flavescens* Behavior. *Horticulturae* **2022**, *8*, 623. https://doi.org/10.3390/horticulturae8070623

Academic Editors: Małgorzata Tartanus and Eligio Malusà

Received: 3 June 2022
Accepted: 7 July 2022
Published: 10 July 2022

Publisher's Note: MDPI stays neutral with regard to jurisdictional claims in published maps and institutional affiliations.

Copyright: © 2022 by the authors. Licensee MDPI, Basel, Switzerland. This article is an open access article distributed under the terms and conditions of the Creative Commons Attribution (CC BY) license (https://creativecommons.org/licenses/by/4.0/).

Abstract: The tea green leafhopper, *Empoasca flavescens*, is a major pest of tea *Camellia sinensis* (L.) Kuntze. Until recently, it has mainly been controlled by pesticides, but their use has led to high levels of toxic residues in plants, which threaten both the environment and human health. Therefore, a safer biological control approach is needed. Tea plants produce many volatile compounds, and different tea clones differ in their resistance to the pest. We explored the possibility that volatile compounds influence the resistance of tea. Here, we assessed the resistance of 15 clones of tea plants to the pest, the volatile compounds produced by the clones, and the effects of the compounds on *E. flavescens* behavior. Six clones were classified as resistant, eight as moderately susceptible, and one as susceptible. Fresh leaf samples from resistant and susceptible clones were analyzed using HS–SPME–GC–MS. Sesquiterpenes and monoterpenes were two major groups characterized, representing 30.15% and 26.98% of the total compounds, respectively. From our analysis, we conclude that 3-hexen-1-ol, 2,6-dimethyleneoct-7-en-3-one, humulene, β-bourbonene, styrene, and benzaldehyde were important for the resistance and susceptibility of the clones. In a bioassay, *E. flavescens* were attracted to β-ocimene and methyl salicylate, but avoided linalool compounds.

Keywords: *Empoasca flavescens*; pest control; tea; volatile compounds

1. Introduction

Tea is a globally popular beverage prepared from the shoots of the tea plant, *Camellia sinensis* (L.) Kuntze, an evergreen shrub native to Asia. Tea is economically important throughout Southeast Asia and is one of Indonesia's most important commercial crops. Tea shoots are delicate, and the plants are susceptible to a variety of insects and pathogens. Of particular concern is the tea green leafhopper, *Empoasca flavescens* (Hemiptera: Cicadellidae), which reduces the tea yield by 15–50% yearly. The adults and nymphs of the leafhopper pierce the shoot and suck the sap from the plant, which causes them to wither [1,2].

Growers have relied on pesticides to control the tea green leafhopper for many decades, but in recent years, as reported in China, the pests have developed resistance to some of the common pesticides, including imidacloprid, bifenthrin, and acetamiprid [3]. The reduced efficacy of these insecticides has caused growers to use more of them. This has led to more toxic residues in the tea and problems in the European tea market, the primary destination of Indonesian tea products. An environmentally sound alternative to control the leafhopper is needed.

Chemicals in the environment are detected by insects and other animals that inhabit the environment. These may function as survival-promoting environments for animals. In particular, an insect's olfactory system is sensitive to volatile compounds that help it in finding habitat, food sources, mating, determining oviposition sites, and escaping from predators [4,5]. Tea plants produce many aromatic compounds that may attract the leafhopper; understanding the bases of these attractions is a potential avenue to control the pest. Different tea clones produce a plethora of volatile chemical substances, which are likely to affect the behavior of insects. For example, tea plants produce (Z)-3-hexenyl acetate and (Z)-3-hexenol, volatiles that induce oviposition in adult females of *E. vitis* [3], and they also produce (E)-ocimene, linalool, (E)-2-hexenal, (Z)-3-hexen-1-ol, (Z)-3-hexenyl acetate, 2-penten-1-ol, (E)-2-pentenal, pentanol, hexanol, and 1-penten-3-ol. The latter group of volatile compounds attracts *E. vitis* [6].

Here, we sought to identify and classify tea clones that are resistant to the tea green leafhopper, identify the major volatiles produced and emitted by tea shoots, and assess the insect's behavioral responses to the volatile compounds produced. The study will provide a foundation for the future development of a system, based on the volatile compounds produced by the tea plant, for monitoring, attracting, and trapping this major tea plantation pest.

2. Materials and Methods

2.1. Experimental Sites

The study was conducted from January to December 2021 at the Indonesian Research Institute for Tea and Cinchona (IRITC) experimental garden in Gambung, Bandung Regency, West Java, Indonesia (7″ 08′37.3″S′ 107′′30′56.3″E). The elevation of the site is 1350 m above sea level. The soil type is Andisols and the pH ranges between 4.5 and 6.0. Schmidt and Ferguson [7] classify the rainfall as category B.

2.2. Plant Materials

The materials used in the study were derived from a cross between TRI 2024 with PS 1 series clones, which were seeded in 1980 (Supplementary Table S1). The TRI 2024 clone is easy to propagate vegetatively, produces numerous shoots, and is of high grade; however, it is susceptible to blister blight disease, has a low shoot weight, and does not respond to nitrogenated fertilizers. Clone PS 1 is resistant to blister blight disease. We performed a cross of these clones to obtain clones with high production and resistance to both biotic and abiotic stresses.

Fifteen clones were tested and compared: I.35.8, II.6.10, II.10.11, II.13.2, II.32.15, III.2.15, III.22.15, III.28.4, III.36.15, TPS 17/3, TPS 24/5, TPS 87/1, TPS 87/2, TPS 93/3, and TPS 122/2. These clones were selected based on root development study, yield potential, and resistance to blister blight [8]. Each clone consisted of 12 plants. The plants were grown with a spacing of 120 × 80 cm. We also used one clone of GMB 7, which was known to be resistant to *E. flavescens* (Supplementary Table S1).

2.3. Resistance Selection in Tea Plants

Resistance tea clones to *E. flavescens* were selected by observing and assessing the population density of *E. flavescens* among the clones grown in the IRITC experimental garden in Gambung. Population observations were conducted using the beat bucket method. We used a randomized block design with a single factor for the experiment and replicated it three times.

2.4. Analysis of Tea Leaf Volatile Compounds

Five grams of each leaf sample from resistant and susceptible clones were harvested from the tea plant in the field between 15:00 and 17:00. The shoots were placed in a 22 mL glass vial fitted with a polytetrafluoroethylene/silicone septum (Agilent). The samples were either two leaves and a bud or three leaves and a bud (Figure 1).

Figure 1. Photograph of the tea shoot sample. The samples taken for the analysis are indicated.

The volatile compounds were extracted from the leaves and analyzed with a solid-phase microextraction (SPME)-headspace (HS)/gas chromatograph–mass spectrometer (GC–MS) (Lin et al., 2012). Tissue samples were extracted at 30 °C for 45 min with an SPME fiber coated with divinylbenzene/carboxen/polydimethylsiloxane (DVB/CAR/PDMS). The fiber was immediately inserted into the GC/MS (Agilent 7890A and Agilent 5975C XL EI/CI instruments, respectively) with splitless column effluent. The metabolites were separated using an HP-5MS column (30 m × 250 m × 0.25 m) with helium as the carrier gas, at a flow rate of 1 mL/min.

The oven temperature was programmed from 50 to 250 °C, and the total working time of the GC's entire operation was 45 min. The MS transfer was conducted at 280 °C, the scanned mass ranged from 29 to 550 amu, the MS source was set to 230 °C, and the MS Quad was set to 150 °C. The oven temperature was held at 50 °C for 5 min and then increased to 220 °C at a rate of 3 °C/min. The retention times of the compounds were compared to those of authentic standards and their mass spectra, as well as to relevant data in the NIST14 Mixture Property Database. The compounds were quantified by comparing their GC total ion current peak areas to the internal standard peak area. The profile of the volatile compounds was analyzed with the principal component analysis (PCA), partial least squares-discriminant analysis (PLS-DA), and heatmap clustering (MetaboAnalyst version 5.0).

2.5. Y-Tube Olfactometer Test

The behavioral responses of *E. flavescens* adults were evaluated to odors released from synthetic compounds in a Y-tube olfactometer (10 × 10 × 10 cm arm length, 3 cm diameter, 75° Y angle) (Supplementary Video S1). These bioassays were conducted between 08:00 and 11:00 or 15:00 and 17:00. Adult leafhoppers were collected randomly from the tea plants using the beat bucket technique, placed in cages (30 × 30 × 30 cm), and fed with tea shoots. For the bioassay, the insects were acclimated by transferring them to test tubes without food for an hour.

Insect responses were tested to the identified compounds, specifically β-ocimene, linalool, and methyl salicylate (Sigma-Aldrich Inc., Steinheim, Germany) with liquid paraffin as the control [2,6]. The volatile compounds were dissolved in liquid paraffin (1%), and 20 µL of solution or paraffin alone was applied to a filter paper square (1 × 1 cm). One arm of the olfactometer [6,9] received a filter paper with the sample and the other arm received paraffin only. Charcoal and humidified, filtered air was driven into each tube with a vacuum pump. After acclimation, individual leafhoppers were placed at the downwind end of the main tube. The insect's 'choice' for either the control or compound was assessed based on when it crossed a line 3 cm past the fork of the base tube and remained there for at least 3 min. If a leafhopper did not choose within 10 min, it was considered unresponsive, and its activity was recorded as 'no choice.' After testing two leafhoppers, the olfactometer tube was rinsed with 70% ethanol, heated at 100 °C for 5 min, and the sample in each arm of the tube was reversed. At least 30 replicates were performed for each volatile compound tested [3].

2.6. Statistical Analysis

The analysis of variance of the population data was used to determine the significance of differences in the mean size of the pest populations in different clones. The analysis of variance was continued with Duncan's multiple range test (DMRT) to group clones with similar mean sizes of pest populations. The results of this statistical analysis served as criteria to categorize the level of tea clone resistance to the leafhoppers. A large pest population associated with a clone indicates that the clone is susceptible to the pest [10]. The behavioral responses were analyzed using a non-parametric statistical binomial test. Data analyses were performed with SPSS 16.0 (SPSS Inc., Chicago, IL, USA).

3. Results

3.1. Resistance Selection of Tea Plant Resistance

The population density of a pest associated with the host plant is an important factor in the damage it can cause to the host. Here, the population density of *E. flavescens* was used as a parameter to assess the resistance of the different tea clones to it. The population density of *E. flavescens* was quite variable for the different clones, as shown in Figure 2, and ranged from 1.63 to 6.99 individuals per plant (Figure 2). We conclude that resistance to *E. flavescens* infestation in the field varied significantly across the 15 TPS clones tested.

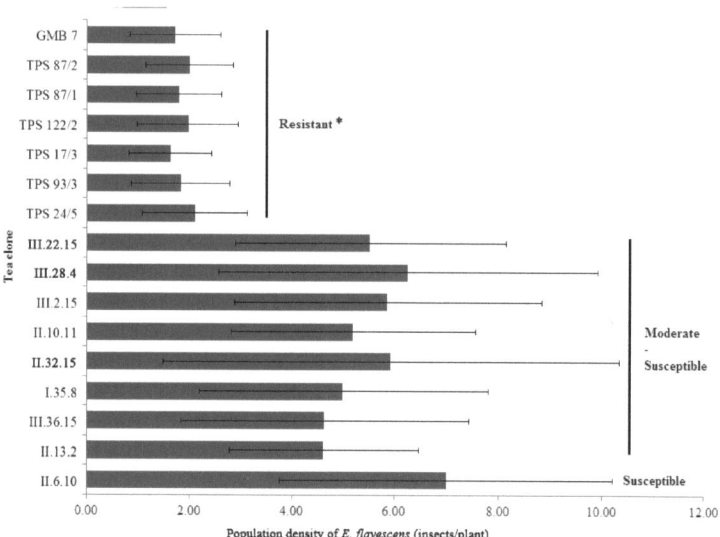

Figure 2. Population density of the leafhopper *E. flavescens* on plants of different tea clones. Data represent mean ± SD. The asterisk indicates significant differences between groups ($p < 0.05$).

The resistance level of the clones was assessed by comparing the population density of each clone to the known resistant clone GMB 7. Using this comparison, we conclude that the clones TPS 87/2, TPS 87/1, TPS 122/2, TPS 17/3, TPS 93/3, and TPS 24/5 are resistant clones. The clones with a population density of 4.62 to 6.99 individuals/plant were considered susceptible. Clones III.22.15, III.28.4, III.2.15, II.10.11, II.32.15, I.35.8, III.36.15, and II.13.2 were classified as moderately susceptible; only clone II.6.10 was susceptible.

3.2. Profile of Volatile Compounds of Resistant and Susceptible Tea Clones

Sixty-three volatile compounds were detected through the GC–MS analysis. The susceptible and moderately susceptible clones produced 53 different volatile compounds, while the resistant clones produced 47 different compounds. These compounds were classified into 11 groups i.e., alcohols, aldehydes or alkanes, ketones, carboxylic acids, esters, aromatics, monoterpenes, sesquiterpenes, alkaloids, fatty acids, sterols, and phenols (Table 1). Many of the compounds were mono- and sesquiterpenes; of the 63 compounds detected, 30.15% were monoterpenes and 26.98% sesquiterpenes.

PCA was used to identify the volatile compounds that conferred the characteristics of resistance and susceptibility to the clones. From the analysis, we can determine which characteristics contributed to the overall diversity of volatile chemicals generated. Three principal components (PCs) were used to represent the original volatile compounds, with cumulative variance levels of 74.7% for PC1 (39%), PC2 (19.4%), and PC3 (16.3%) (Figure 3A); we interpret this to mean that the combination of the two PCs accounted for 74.7% of the information obtained from the original. From the PCA analysis, we classified the compounds as belonging to two distinct groups: the resistant clone group (TPS 24/5; TPS 122/2; and TPS 93/3) or the susceptible clone group (II.10.11; II.13.2; and II.06.10). Some compounds were present in both groups; these were detected in the susceptible clone II.12.3 and the resistant clone TPS 93/3 (Figure 3B).

Table 1. The volatile compounds produced by the tea clones and the chemical groups they belong to. Compounds that are important in the resistance or susceptibility of the clones to the leafhopper *E. flavescens* are listed.

No	RT (min)	Compound	Area (%) Susceptible	Area (%) Resistant	Group
1	1.4828	acetaldehyde	16.61	19.35	Aldehyde/alkane
2	2.7629	cyclobutanol	4.41	5.03	Alcohol
3	3.7990	hexanal	0.53	1.39	Aldehyde/alkane
4	5.4946	3-hexen-1-ol. (E)-	1.33	-	Alcohol
5	6.6175	styrene	-	1.12	Aromatic
6	9.3340	benzaldehyde	-	0.18	Aromatic
7	10.2260	1-octen-3-ol	0.32	2.72	Alcohol
8	10.6340	β-myrcene	2.58	2.59	Monoterpene
9	12.0731	D-limonene	0.92	0.80	Monoterpene
10	12.5249	trans-β-ocimene	1.17	0.90	Monoterpene
11	12.9299	β-ocimene	10.18	5.24	Monoterpene
12	13.8942	trans-linalool oxide	1.09	0.82	Monoterpene
13	14.4410	terpinolene	0.05	0.05	Monoterpene
14	14.5413	cis-linalool oxide	2.23	1.90	Monoterpene
15	14.7960	4.8-dimethyl-1.3.7-nonatriene	0.35	0.22	Alcohol
16	15.2901	linalool	38.86	32.29	Monoterpene
17	15.6572	2.6-dimethyleneoct-7-en-3-one	-	0.68	Ketone
18	16.1447	allo-ocimene	0.86	0.61	Monoterpene
19	16.6140	neo-allo-ocimene	0.56	0.42	Monoterpene
20	17.1008	2.6-nonadienal. (E.Z)-	0.06	0.08	Aldehyde/alkane
21	17.5362	isoborneol	-	0.06	Monoterpene
22	17.7293	ethyl benzoate	0.10	0.08	Aromatic
23	17.9429	epoxylinalol	0.22	0.20	Monoterpene
24	18.0351	napthalene	0.05	-	Aromatic
25	18.7876	methyl salicylate	14.88	20.00	Carboxylic acid
26	18.8602	2-ethoxybenzoic acid	0.13	-	Carboxylic acid
27	19.2514	methyl aspirin	0.02	0.03	Carboxylic acid
28	19.8077	2-carene	0.03	-	Monoterpene
29	19.9878	cis-3-hexenyl-alpha-methylbutyrate	0.09	0.06	Ester
30	20.1350	cis-3-hexenyl valerate	0.11	0.11	Ester

Table 1. Cont.

No	RT (min)	Compound	Area (%) Susceptible	Area (%) Resistant	Group
31	20.3244	hexyl valerate	0.02	0.02	Ester
32	20.4539	isogeraniol	0.09	-	Monoterpene
33	20.8522	geraniol	0.15	0.41	Monoterpene
34	21.3253	ethyl salicylate	0.64	0.63	Carboxylic acid
35	22.2461	indole	0.12	0.18	Aromatic
36	22.8427	3-ethoxy-2-pyridinamine	0.02	-	Alkaloid
37	22.9760	methyl m-methoxymandelate	0.05	0.04	Aromatic
38	23.2036	octane. 2.4.6-trimethyl-	0.02	-	Fatty acid
39	25.1588	4-ethoxybenzaldehyde	-	0.17	Phenol
40	25.1649	β-bourbonene	-	0.01	Sesquiterpene
41	25.4249	β-elemene	-	0.03	Sesquiterpene
42	25.6863	cis-jasmone	0.04	-	Ketone
43	25.8880	isocaryophyllene	0.04	-	Sesquiterpene
44	26.0422	ledene	-	0.04	Sesquiterpene
45	26.2873	caryophyllene	0.49	0.79	Sesquiterpene
46	26.3850	cholestan-2-one oxime	0.01	0.01	Sterol
47	26.6687	β-cubebene	-	0.03	Sesquiterpene
48	27.2546	aromandendrene	0.06	0.03	Sesquiterpene
49	27.3715	humulene	-	0.20	Sesquiterpene
50	27.4949	(E)-β-farnesene	0.01	0.04	Sesquiterpene
51	27.6045	alloaromadendrene	0.01	0.04	Sesquiterpene
52	28.1071	γ-muurolene	0.01	-	Sesquiterpene
53	28.6874	longifolene	0.05	-	Sesquiterpene
54	29.0986	α-farnesene	0.23	0.07	Sesquiterpene
55	29.2642	θ-muurolene	0.01	-	Sesquiterpene
56	29.5492	calamenene	0.06	0.15	Sesquiterpene
57	30.1420	α-calacorene	0.01	-	Sesquiterpene
58	30.6387	trans-farnesol	0.01	-	Sesquiterpene
59	30.7600	E-nerolidol	0.02	0.15	Sesquiterpene
60	30.9116	6-octenal. 7-methyl-3-methylene-	0.01	-	Aldehyde/alkane
61	31.1558	(Z.E)-farnesol	0.04	0.02	Sesquiterpene
62	31.6900	7.9-dimethylhexadecane	0.01	0.01	Fatty acid
63	32.2819	α-patchoulene	0.02	-	Sesquiterpene

From the PLS-DA analysis, we identified the volatile compounds that distinguished resistant from susceptible clones (Figure 3C). Two major components were apparent, with a total variance of 79.3% (PC1 74.7% and PC3 4.9%). In general, the PC value describes how clones are grouped based on the compounds produced. Variable importance for the projection (VIP) scores greater than 1 identify important volatile substances within groups [1]. Fifteen compounds were detected, specifically 3-hexen-1-ol; 1-octen-3-ol; styrene; methyl salicylate; 2,6-dimethyleneoct-7-en-3-one; hexanal; β-ocimene; humulene; β-bourbonene; linalool; cis-linalool oxide; trans-linalool oxide; caryophyllene; α-farnesene, and benzaldehyde (Figure 3D). The color indicator in Figure 3D shows the concentration of the chemical in each clone. Dark red denotes a very high concentration, whereas dark blue indicates a very low concentration.

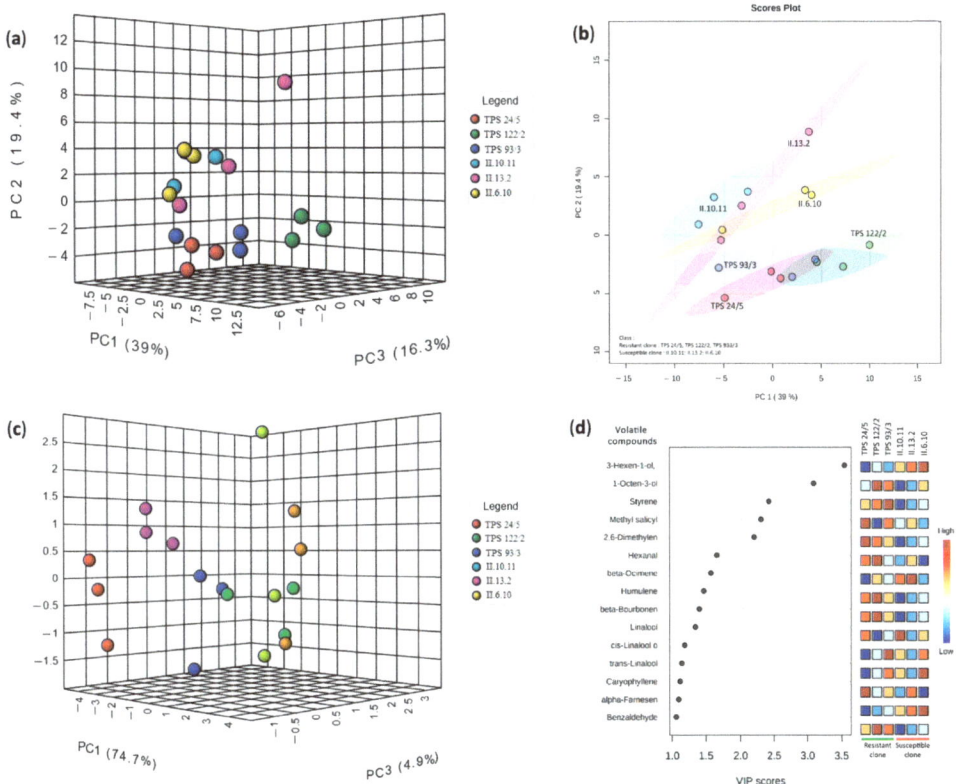

Figure 3. Principal component analysis (PCA) of the volatile compounds that conferred the characteristics of resistance and susceptibility of the tea clones to the leafhopper *E. flavescens*. (**a**) *Scree plot* PCA; (**b**) *Scores plot* PCA; (**c**) *Scores plot* (partial least squares-discriminant analysis (PLS-DA); (**d**) variable importance for the projection (VIP) *scores* PLS-DA. (PC, principal component).

A heatmap was constructed to visualize the compounds that may confer resistance and susceptibility. This highlights the variations in the chemicals generated by each group of clones (Figure 4). Resistant clones produced several compounds absent in the susceptible clones, including 2,6-dimethyleneoct-7-en-3-one; humulene; β-bourbonene; styrene; 1-octen-3-ol; and benzaldehyde. In contrast, the susceptible clones produced 3-hexen-1-ol, which was not detected in the resistant clones. The compounds used to test the response of *E. flavescens* in the bioassay were β-ocimene, linalool, and methyl salicylate; these compounds are produced by all clones and are dominant.

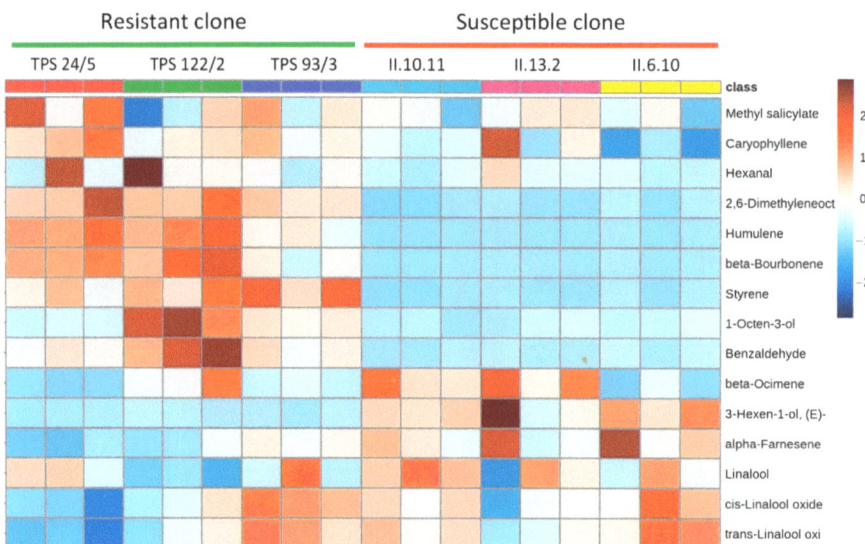

Figure 4. Heatmap cluster based on the normalized quantities of the volatile compounds identified in the tea clones. The color indicators indicate relative amounts of the compounds. The darkest blue indicates the least amount of a compound and the darkest red the most.

3.3. The Response of E. flavescens Response to the Volatile Compounds Produced by Tea Plants

The significance of the preference of *E. flavescens* on the treatment was analyzed using the binomial test. *E. flavescens* showed attraction to methyl salicylate and β-ocimene, and tended to avoid linalool (Figure 5).

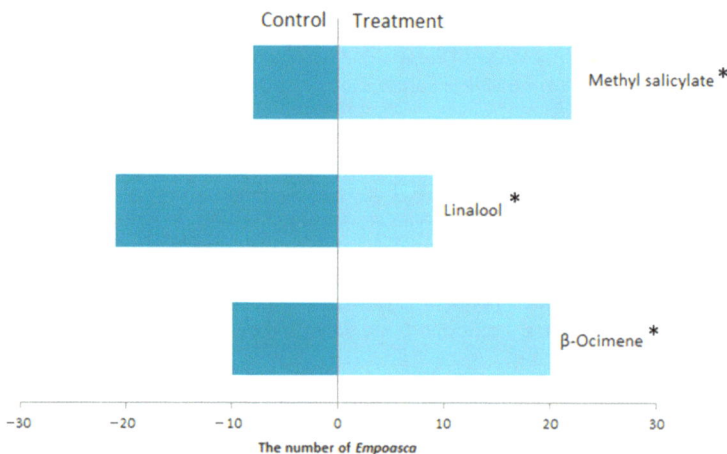

Figure 5. Bioassay results of *E. flavescens* to methyl salicylate, linalool, and β-ocimene. The insects either selected or avoided the region of a Y-tube olfactometer that contained the volatile compound. The asterisk indicates significant differences between groups (two-tailed binomial test).

4. Discussion

We assessed the population density of *E. flavescens* associated with each of the 15 TPS clones we studied. From the results, we conclude that the responses of the clones to the pests were variable. The clones were classified based on the population densities of the pests associated with them in the field. Susceptible clones had significantly higher densities than those clones classified as resistant. From the established criteria, six clones were resistant, eight clones were moderately susceptible, and one clone was susceptible. Plant resistance to insect pests can be defined as the ability of plants to withstand pest attacks and minimize damage. Clones that are resistant to a pest generally suppress the development of the pest population, preventing economic losses from the infestation [11].

Plants have evolved various defense mechanisms to protect themselves from insect pests. Some emit volatile compounds that have direct effects on herbivores. These vary between different types of plants and may vary in the same plant in different situations. Plants use volatile compounds as indirect defenses as well. For example, a plant under attack by an herbivore may emit volatile pheromones that attract insect predators of the attacking herbivore. When "summoned" by the plant in this way, the predator destroys the initial pest; the plant's volatiles may also communicate danger signals to other plants [12]. When tea plants are attacked by their common pests, they may emit numerous herbivore-induced plant volatiles, such as (Z)-3-hexenol, linalool, α-farnesene, benzyl nitrile, indole, nerolidol, and ocimenes at high concentrations [13].

Xin, Li, Bian, and Sun [3] studied the volatile compounds produced in the field by tea varieties that differed in resistance and susceptibility to *E. vitis*. They compared two resistant varieties, Changxingzisun and Juyan, with two susceptible varieties, Enbiao and Banzhuyuan. The population density of *E. vitis* associated with the two susceptible varieties was greater than that associated with the resistant varieties, confirming their classification as susceptible and resistant. They also measured the emission levels of (Z)-3-hexenyl in all four varieties. More of this volatile compound was produced by the susceptible varieties Enbiao and Banzhuyuan than by the two resistant varieties. The plant variety is one of the main factors that affect the composition of the volatile organic compound blend released by plants. This background inspired us to characterize the characteristics of volatile compounds emitted by resistant and susceptible tea clones.

GC–MS was used to analyze the volatile compounds produced by the tea clones that have been selected. Most of the detected volatiles were terpenoids, either monoterpenes or sesquiterpenes. In this regard, our results are similar to those of Lin, et al. [14], who observed that the volatile compounds in Longjing tea were dominated by sesquiterpenes. In addition to sesquiterpenes, others have detected monoterpenes, alcoholic terpenes, alcohols, esters, aldehydes, ketones, and alkanes in tea plants [14]. Terpenes in tea plants are important in biological processes unrelated to plant defense and are the main components in the aroma; indeed, the particular terpenes present are determinants of the tea quality [15,16]. Most of the distinctive aromatic compounds in tea are terpenes, specifically geraniol, farnesene, ocimene, linalool, and nerolidol [16]. Monoterpene compounds, such as linalool and sesquiterpene (E)-β-farnesene, have insect-repellent properties [15].

Each of the 15 clones in this study produced a different set of volatile compounds, which may be a distinguishing feature of that clone. The susceptible clones produced 3-hexen-1-ol, which is commonly found in fresh tea leaves. Its concentration increases when the plant is mechanically injured or attacked by pests. Xin, Li, Bian, and Sun [3] reported that adult male and female *E. vitis* were attracted to 3-hexen-1-ol compounds, and the females preferred the susceptible varieties Enbiao and Banzhuyuan for egg-laying over the resistant varieties. The susceptible varieties produced 3-hexen-1-ol and (Z)-3-hexenyl acetate compounds, while in the resistant varieties, these compounds were detected in much lower amounts. These investigators suggested that these two compounds are signals for adult female insects to detect and choose the location of oviposition.

The resistant clones in this study produced more compounds than the susceptible clones, specifically, humulene, 2,6-dimethyloct-7-en-3-one, β-bourbonene, styrene, and benzaldehyde. Humulene is important in plant resistance to pests. It has antimicrobial, anti-tumor, anti-fungal, and anti-inflammatory properties and is the main component of essential oils from various types of plants. The concentration of humulene varies in different plants but is part of the essential oils of many aromatic plants, notably: *Salvia officinalis, Lindera strychnifolia* Uyaku, ginseng, *Mentha spicata*, ginger, *Litsea mushaensis, Cordia verbenacea*, Vietnamese coriander, *Humulus lupulus*, pine trees, citrus, tobacco, and sunflowers [17,18]. Humulene has been developed as a biopesticide for several types of pests. For example, extracts of *Zingiber officinale, Curcuma longa*, and *Alpinia galanga* contain humulen and are effective for controlling the warehouse pests *Sitophilus zeamais* and *Tribolium castaneum*. The compound also inhibits the development of *Conopomorpha cramerella* Snellen pupae by as much as 49.75% compared to untreated pupae [18]. Humulene essential oil from leaves of *Commiphora leptophloeos* suppressed the oviposition of the *Aedes aegypti* mosquito by as much as 31.2% compared to the control [17]. Based on the results of this study, it may be that humulene is important in the defense system of resistant clones of tea to *E. flavescens*.

The volatile compound 1-octen-3-ol is known as a "fungal alcohol" and is usually associated with the fungi *Aspergillus, Cladosporium, Mucor*, and *Ulocladium*. It was produced by both resistant and susceptible clones. Detection of the aroma of 1-octen-3-ol can be an indicator of mold in an area. Application of 1-octen-3-ol suppressed oviposition of *Drosophila suzukii* on raspberry commodities by up to 55%, impaired locomotion, and caused lipid peroxidation, slow development, cell apoptosis, neurotoxicity, and inflammation [19,20]. Application of 98 L/L 1-octen-3-ol as a fumigant was 100% effective in eradicating the warehouse pest *Tribolium castaneum* and negatively affected the development and reproduction of adult *T. castaneum* and its progeny, resulting in decreased survival, reduced pupae weight, and reduced adult insect weight [21].

Beta-ocimene, linalool, and methyl salicylate were produced in the greatest amounts by the 15 clones in this study, compared to the other compounds detected. This was true for both resistant and susceptible clones. In insects, β-ocimene may both attract and repel. In this study, *E. flavescens* were attracted to methyl salicylate and β-ocimene. In another study, *E. vitis* and *E. onukii* were both attracted to (E)-ocimene in a laboratory bioassay using a Y-tube olfactometer [4,6]. *Pseudotheraptus wayi* recognized its host plant (the cashew) by detecting the ocimene compound, while its natural enemy, *Oecophylla longinoda* (Latreille) was also attracted to ocimene [22].

This research showed that *E. flavescens* was attracted not only to β-ocimene, but also to methyl salicylate. Methyl salicylate released by arthropods was attractive to both beneficial insects and insect pests. Methyl salicylate applied to yellow sticky traps attracts herbivores, including leafhoppers, plant bugs (Miridae), weevils, thrips, and soldier beetles in cranberries [23]. Methyl salicylate is also important for the beetle *Anthonomus grandis* to detect cotton plants, and it was the most effective attractant studied for the spotted lanternfly [24,25]. Linalool is the predominant compound in both fresh and prepared tea leaves. In this study, *E. flavescens* avoided linalool. Linalool can act as a repellent to female insects and suppress oviposition events and the number of eggs laid by female *Ceratitis capitata* on citrus fruits [26]. Linalool was also repellent to aphids, Lepidoptera, and thrips in transgenic *Arabidopsis* and *Chrysanthemum* plants that overexpressed the linalool synthesis gene [26]. We suggest that linalool might be a promising compound for pest control. The behavioral response shown by *E. flavescens* to volatile compounds needs to be studied further to determine the response at different doses and mixtures of compounds. Field testing is also needed to identify what risks are posed by these natural pesticides in the environment.

5. Conclusions

We conclude that the volatile compounds β-ocimene, linalool, and methyl salicylate tested have excellent potential to control *E. flavescens* in tea plants. This approach is environmentally friendly and should lead to reduced use of herbicides.

Supplementary Materials: The following supporting information can be downloaded at: https://www.mdpi.com/article/10.3390/horticulturae8070623/s1, Table S1: Tea clones description; Video S1: Olfactometer test.

Author Contributions: Conceptualization, methodology, investigation, F.F., A.D.P. and A.F.; data curation, F.F.; writing—original draft preparation, F.F.; writing—review and editing, A.F. and A.D.P.; supervision, A.F. and A.D.P.; project administration, A.F.; funding acquisition, A.F. All authors have read and agreed to the published version of the manuscript.

Funding: This research was partially funded by the Ministry of Education, Culture, Research, Technology, and Higher Education, the Republic of Indonesia, under the scheme of excellent research for a university grant, granted to the first author (contract no. 2/E1/KP.PTNBH/2021) and we thank the Educational Fund Management Institution, Ministry of Finance, the Republic of Indonesia for providing financial support for F.F.

Institutional Review Board Statement: Not applicable.

Informed Consent Statement: Not applicable.

Data Availability Statement: Not applicable.

Conflicts of Interest: The authors declare no conflict of interest.

References

1. Mu, D.; Pan, C.; Qi, Z.; Qin, H.; Li, Q.; Liang, K.; Rao, Y.; Sun, T. Multivariate analysis of volatile profiles in tea plant infested by tea green leafhopper *Empoasca onukii* Matsuda. *Plant Growth Regul.* **2021**, *95*, 111–120. [CrossRef]
2. Zhang, Z.; Chen, Z. Non-host plant essential oil volatiles with potential for a 'push-pull' strategy to control the tea green leafhopper, *Empoasca vitis*. *Entomol. Exp.* **2015**, *156*, 77–87. [CrossRef]
3. Xin, Z.J.; Li, X.W.; Bian, L.; Sun, X.L. Tea green leafhopper, *Empoasca vitis*, chooses suitable host plants by detecting the emission level of (3Z)-hexenyl acetate. *Bull. Entomol. Res.* **2017**, *107*, 77–84. [CrossRef] [PubMed]
4. Zhao, Y.; Li, H.; Wang, Q.; Liu, J.; Zhang, L.; Mu, W.; Xu, Y.; Zhang, Z.; Gu, S. Identification and expression analysis of chemosensory genes in the tea green leafhopper, *Empoasca onukii* Matsuda. *J. Appl. Entomol.* **2018**, *142*, 828–846. [CrossRef]
5. Fleischer, J.; Pregitzer, P.; Breer, H.; Krieger, J. Access to the odor world: Olfactory receptors and their role for signal transduction in insects. *Cell. Mol. Life Sci.* **2018**, *75*, 485–508. [CrossRef]
6. Mu, D.; Cui, L.; Ge, J.; Wang, M.-X.; Liu, L.-F.; Yu, X.-P.; Zhang, Q.-H.; Han, B.-Y. Behavioral responses for evaluating the attractiveness of specific tea shoot volatiles to the tea green leafhopper, *Empoaca vitis*. *Insect Sci.* **2012**, *19*, 229–238. [CrossRef]
7. Schmidt, F.H.; Ferguson, J.H.A. *Rainfall Types Based on Wet and Dry Period Ratios for Indonesia with Western New Guinee*; Kementerian Perhubungan: Jakarta, Indonesia, 1952.
8. Sriyadi. Seleksi ketahanan klon teh seri TPS terhadap penyakit cacar. *J. Penelit. Teh. Dan Kina* **2007**, *10*, 73–82.
9. Cai, X.; Luo, Z.; Meng, J.; Liu, Y.; Chu, B.; Bian, L.; Li, Z.; Xin, Z.; Chen, Z. Primary screening and application of repellent plant volatiles to control tea leafhopper, *Empoasca onukii* Matsuda. *Pest Manag. Sci.* **2020**, *76*, 1304–1312. [CrossRef]
10. Wagiman, F.X.; Triman, B. Ketahanan relatif enam belas nomor klon teh PGL terhadap serangan *Empoasca* sp. *J. Perlindungan Tanam. Indones.* **2011**, *17*, 60–65.
11. Zhou, S.; Jander, G. Molecular ecology of plant volatiles in interactions with insect herbivores. *J. Exp. Bot.* **2021**, *73*, 449–462. [CrossRef]
12. Chen, C.; Chen, H.; Huang, S.; Jiang, T.; Wang, C.; Tao, Z.; He, C.; Tang, Q.; Li, P. Volatile DMNT directly protects plants against *Plutella xylostella* by disrupting the peritrophic matrix barrier in insect midgut. *eLife* **2021**, *10*, e63938. [CrossRef] [PubMed]
13. Jing, T.; Du, W.; Gao, T.; Wu, Y.; Zhang, N.; Zhao, M.; Jin, J.; Wang, J.; Schwab, W.; Wan, X.; et al. Herbivore-induced DMNT catalyzed by CYP82D47 plays an important role in the induction of JA-dependent herbivore resistance of neighboring tea plants. *Plant Cell Environ.* **2021**, *44*, 1178–1191. [CrossRef] [PubMed]
14. Lin, J.; Dai, Y.; Guo, Y.-N.; Xu, H.-R.; Wang, X.-C. Volatile profile analysis and quality prediction of Longjing tea (*Camellia sinensis*) by HS-SPME/GC-MS. *J. Zhejiang Univ. Sci. B* **2012**, *13*, 972–980. [CrossRef]
15. Mithöfer, A.; Boland, W. Plant defense against herbivores: Chemical aspects. *Annu. Rev. Plant Biol.* **2012**, *63*, 431–450. [CrossRef]
16. Zeng, L.; Watanabe, N.; Yang, Z. Understanding the biosynthesis and stress response mechanisms of aroma compounds in tea (*Camellia sinensis*) to safely and effectively improve tea aroma. *Crit. Rev. Food Sci. Nutr.* **2019**, *59*, 2321–2334. [CrossRef] [PubMed]
17. da Silva, R.C.S.; Milet-Pinheiro, P.; Bezerra da Silva, P.C.; da Silva, A.G.; da Silva, M.V.; Navarro, D.M.d.A.F.; da Silva, N.H. (E)-caryophyllene and α-humulene: *Aedes aegypti* oviposition deterrents elucidated by gas chromatography-electrophysiological assay of *Commiphora leptophloeos* leaf oil. *PLoS ONE* **2015**, *10*, e0144586. [CrossRef] [PubMed]
18. Ninkuu, V.; Zhang, L.; Yan, J.; Fu, Z.; Yang, T.; Zeng, H. Biochemistry of terpenes and recent advances in plant protection. *Int. J. Mol. Sci.* **2021**, *22*, 5710. [CrossRef]
19. Macedo, G.E.; de Brum Vieira, P.; Rodrigues, N.R.; Gomes, K.K.; Martins, I.K.; Franco, J.L.; Posser, T. Fungal compound 1-octen-3-ol induces mitochondrial morphological alterations and respiration dysfunctions in *Drosophila melanogaster*. *Ecotoxicol. Environ. Saf.* **2020**, *206*, 111232. [CrossRef]
20. Stockton, D.G.; Wallingford, A.K.; Cha, D.H.; Loeb, G.M. Automated aerosol puffers effectively deliver 1-octen-3-ol, an oviposition antagonist useful against spotted-wing drosophila. *Pest Manag. Sci.* **2021**, *77*, 389–396. [CrossRef]

21. Cui, K.; Zhang, L.; He, L.; Zhang, Z.; Zhang, T.; Mu, W.; Lin, J.; Liu, F. Toxicological effects of the fungal volatile compound 1-octen-3-ol against the red flour beetle, *Tribolium castaneum* (Herbst). *Ecotoxicol. Environ. Saf.* **2021**, *208*, 111597. [CrossRef]
22. Vasconcelos, J.F.; Dias-Pini, N.; Saraiva, W.V.A.; Farias, L.D.L.; Ribeiro, P.R.V.; Melo, J.W.D.S.; Rodrigues, T.H.S.; Macedo, V.H.M. Volatile and phenolic compounds In the resistance of the melon to the vegetable leafminer, *Liriomyza sativae* Blanchard (Diptera: Agromyzidae). *J. Chem. Ecol.* **2022**, 1–22. [CrossRef]
23. Salamanca, J.; Souza, B.; Kyryczenko-Roth, V.; Rodriguez-Saona, C. Methyl salicylate increases attraction and function of beneficial arthropods in cranberries. *Insects* **2019**, *10*, 423. [CrossRef] [PubMed]
24. Derstine, N.T.; Meier, L.; Canlas, I.; Murman, K.; Cannon, S.; Carrillo, D.; Wallace, M.; Cooperband, M.F. Plant volatiles help mediate host plant selection and attraction of the spotted lanternfly (Hemiptera: Fulgoridae): A generalist with a preferred host. *Environ. Entomol.* **2020**, *49*, 1049–1062. [CrossRef] [PubMed]
25. Magalhães, D.M.; Borges, M.; Laumann, R.A.; Woodcock, C.M.; Withall, D.M.; Pickett, J.A.; Birkett, M.A.; Blassioli-Moraes, M.C. Identification of volatile compounds involved in host location by *Anthonomus grandis* (Coleoptera: Curculionidae). *Front. Ecol. Evol.* **2018**, *6*, 98. [CrossRef]
26. Papanastasiou, S.A.; Ioannou, C.S.; Papadopoulos, N.T. Oviposition-deterrent effect of linalool—A compound of citrus essential oils—On female Mediterranean fruit flies, *Ceratitis capitata* (Diptera: Tephritidae). *Pest Manag. Sci.* **2020**, *76*, 3066–3077. [CrossRef]

Article

Production of *Bacillus velezensis* Strain GB1 as a Biocontrol Agent and Its Impact on *Bemisia tabaci* by Inducing Systemic Resistance in a Squash Plant

Ahmed Soliman [1], Saleh Matar [2,3] and Gaber Abo-Zaid [3,*]

1. Applied Entomology and Zoology Department, Faculty of Agriculture (EL-Shatby), Alexandria University, Alexandria 21545, Egypt; soliman.m.ahmad10@gmail.com
2. Chemical Engineering Department, Faculty of Engineering, Jazan University, Jazan 45142, Saudi Arabia; salehmatar@yahoo.com
3. Bioprocess Development Department, Genetic Engineering and Biotechnology Research Institute (GEBRI), City of Scientific Research and Technological Applications (SRTA-City), New Borg El-Arab City, Alexandria 21934, Egypt
* Correspondence: gaberam57@yahoo.com; Tel.: +20-1226804278

Abstract: Pests represent a huge problem in crop production causing significant losses. Currently, biocontrol is utilized as an eco-friendly approach for controlling pests and reducing the shortage in crop production. In the current study, the production of a biocontrol agent, which was identified based on sequencing of the 16S rRNA gene as *Bacillus velezensis* strain GB1 with GenBank accession No. OM836750, was carried out in the stirred tank bioreactor using a batch fermentation process. For the first time, *B. velezensis* strain GB1 was tested as a biocontrol agent with soil drench application (10^9 cfu mL^{-1}) for management of *Bemisia tabaci* and induction of squash plant systemic resistance under greenhouse conditions. β-1,3-glucanase, chitinase, polyphenol oxidase, and peroxidase activity were measured in squash leaves at 24, 48, 72, 96, and 120 h. The influence of *B. velezensis* strain GB1 on population density, fertility, and hatchability of *B. tabaci* on squash plants was studied. The batch fermentation process of *B. velezensis* strain GB1 maximized the production of secondary metabolites and culture biomass, which reached a maximum value of 3.8 g L^{-1} at 10.5 h with a yield coefficient of 0.65 g cells/g glucose. Treatment with *B. velezensis* strain GB1 induced squash plants to boost their levels of β-1,3-glucanase, chitinase, polyphenol oxidase, and peroxidase enzymes. On the other hand, *B. velezensis* strain GB1 could significantly reduce the mean number of the attracted *B. tabaci* on squash plants. Additionally, whiteflies laid a lower mean number of 2.28 eggs/female/day on squash plants inoculated with *B. velezensis* strain GB1 compared to control. The percentage of *B. tabaci* egg hatchability declined by 5.7% in the *B. velezensis*-inoculated squash plants.

Keywords: *Bemisia tabaci*; *Bacillus velezensis*; biocontrol agent; fermentation process; pathogenesis-related (PR) proteins; induced systemic resistance

1. Introduction

Bemisia tabaci (Gennadius) (Homoptera: Sternorrhyncha: Aleyrodidae) is a major pest to ornamental plants and horticultural crops, in both the field and greenhouses around the world. Over 600 plants have been identified as hosts for this common pest [1]. A heavy *B. tabaci* infestation can result in decreased plant vigor as well as a variety of physiological disorders [2,3]. Furthermore, the development of nymphs is commonly linked to the formation of sooty molds, which limit photosynthesis and cause defoliation and stunting [4]. Virus transmission can potentially cause significant damage as *B. tabaci* is a vector for over 100 plant viruses [5–7]. Tomato yellow leaf curl virus (TYLCV) and cucumber vein yellowing virus (CVYV) are two examples. These viruses are a severe problem for tomato and cucumber plants, causing yield loss of 50 to 100 percent [8,9].

Chemical pesticides are frequently employed to manage *B. tabaci* because of their instantaneous action, but this method has various downsides, including food safety concerns, insecticide resistance, ecological hazard, and non-target organism effects. To effectively control *B. tabaci*, biocontrol agents have been developed as a substitute for the traditional use of chemical pesticides in an integrated pest management (IPM) system [10].

Induction of plant resistance against the herbivore, which is a new biological strategy for dealing with plant stress conditions could be investigated as a possible approach for managing whitefly infestation [11]. Many investigations revealed that one of the plant defense mechanisms that defend against pathogens or insects attack is increasing the concentrations of the secondary metabolites or some of the host proteins, several of which are referred to as pathogenesis-related (PR) proteins. They currently comprise 17 families of stress proteins including β-1,3-glucanases, chitinases, and peroxidases [12,13]. This can change nutritional quality and palatability, raise toxicity, and change the host plant's anatomy, phenology, and physiology. These chemicals have attracted interest due to their possible causal role in resistance as seen by their high induction during induced local and systemic resistance by which it inhibits the growth and spread of such pathogens and insects.

Beneficial bacteria that live freely in the soil and rhizosphere are known as plant growth-promoting rhizobacteria (PGPR) [14,15]. It enhances plant growth and induces systemic resistance, making the plant more resistant to a wide spectrum of pathogen attacks in the future. This long-term systemic resistance induced by PGPR has been termed induced systemic resistance (ISR) [16,17], which is characterized by the formation of a primed state for defense in which defense-related responses are elicited more quickly in response to pathogen or insect attack [18]. ISR has been reported to be used by certain PGPR to protect plants from pathogens infections [19,20]. However, there have been a few types of research on the ISR employed by PGPR to combat insects [21–24].

Genus *Bacillus* is frequently used as a biocontrol agent [9]. It can improve plant growth and supply plant protection in a variety of crops such as cucumber [8,15,25]. *B. velezensis* is a widely distributed aerobic, endospore-forming, and Gram-positive species of *Bacillus*, which was named by Ruiz-García et al. [26]. It has been extensively researched and employed because of its direct or indirect growth promotion effect for many plants. *B. velezensis* has been observed to suppress the growth of a variety of pathogenic fungi, including *Fusarium oxysporum* [27], *Aspergillus flavus* [28], bacteria, and nematodes via the biosynthesis of secondary metabolites, such as β-1,3-1,4-glucanase, lipopeptide antibiotics (surfactin, iturin, and fengycin, for example), and iron carriers [29–31], which play significant roles in trigger systemic resistance in plants [32]. However, its activity against insects attack and underlying cellular and molecular defense mechanisms has not yet been widely elucidated [33].

The main objective of the current study is the production of *B. velezensis* as a biocontrol agent using a batch fermentation process in the stirred tank bioreactor, and evaluating for the first time the effects of squash (*Cucurbita pepo*) root colonization by *B. velezensis* on *B. tabaci* population density, females fecundity, and the egg hatchability under greenhouse conditions by elucidation of its underlying mechanism in terms of enhancing the expression of pathogenesis-related proteins such as enzymes involved in the build-up a defense strategy against whitefly, which could be used through the integrated pest management programs.

2. Materials and Methods

2.1. The Laboratory Culture of B. tabaci

Adults of *B. tabaci* (Gennadius) (Order: Homoptera, Suborder: Sternorrhyncha, Family: Aleyrodidae) were reared on healthy tomato plants *(Lycopersicon esculentum* Miller). The mother colony of whiteflies was established by Prof. El-Helaly and has been reared at the Department of Applied Entomology, Faculty of Agriculture, Alexandria University since the 1960s. Recently, the mother colony of whiteflies was re-identified by Dr. Jon Martin, Insect/Plant Division, Department of Entomology, The Natural History Museum, UK.

2.2. Isolation and Molecular Identification of the Bacterial Isolate

The bacterial isolate used in this study as a biocontrol agent was isolated from the rhizosphere of eggplant roots (Alexandria—Egypt) by Dr. Abo-Zaid, G.A., Bioprocess Development Department, Genetic Engineering and Biotechnology Research Institute (GEBRI), City of Scientific Research and Technological Applications (SRTA-City), Egypt.

Total DNA was extracted from an antagonistic bacterial isolate isolated in the current investigation according to Istock et al. [34]. Universal primers, Start (forward) 5′AGAGTTTGATCMTGGCTCAG 3′ and End (reverse) 5′TACGGYACCTTGTTACGACTT 3′ were used to amplify the whole length of the 16S rRNA gene [35]. The amplified 16S rRNA gene of the antagonistic bacterial isolate was purified and sequenced based on the enzymatic chain terminator approach by the use of a Big Dye terminator sequencing kit. The nucleotide sequences were then compared to other 16S rRNA gene sequences in the GenBank database (http://www.ncbi.nlm.nih.gov) (accessed on 2 March 2022). The phylogenetic tree was built using the Neighbor-Joining method in MEGA software version 5 (SRTA-City), with a total of 2000 bootstrap replications.

2.3. Production of the Bacterial Isolate

2.3.1. Preparation of Stirred Tank Bioreactor

Batch fermentation was carried out in a 10 L bench-top bioreactor (Cleaver, Saratoga, NY, USA) equipped with three six-bladed disc-turbine impellers and four baffles and controlled by a digital control unit with a working volume of 4 L. The process was automated using a control unit with a 10.4-inch color touch-screen interface and the ability to store up to 59,994 distinct programs for various situations. The temperature and pH levels were established at 30 °C and 7, respectively. Automatic feeding of 2 mol L^{-1} NaOH and 2 mol L^{-1} HCl kept the pH in check. The air was compressed and adjusted to 1 VVM (air volume per broth volume per minute) after passing through a sterile filter. The dissolved oxygen level was kept over 20% by changing the agitation speed among 200 and 600 rpm. The dissolved oxygen level and pH values were determined online using METTLER TOLEDO electrodes.

2.3.2. Batch Fermentation Process

The bacterial isolate was pre-cultured by inoculating a single colony of the isolate into a 500 mL Erlenmeyer flask containing 100 mL of Number 3 production medium [36]. The bacterial isolate was grown overnight at 30 °C and shacked at 200 rpm. The batch fermentation process of the bacterial isolate was initiated in the bioreactor using Number 3 production medium and an optical density (O.D$_{550}$) of 0.5. Several samples of culture were taken throughout the fermentation period, and optical density at 550 nm was used as an indicator of cell number.

2.3.3. Biomass Estimation

Dry cell weight was determined using a 10 mL sample of culture broth, which was centrifuged at 894× g for 10 min, and the pellet was resuspended, washed, and centrifuged again as before. After that, drying the pellets was performed overnight in a dry-air oven at 80 °C [37].

2.3.4. Glucose Estimation

An enzymatic colorimetric kit (Diamond Diagnostics, Egypt) was used for measuring glucose concentration, which is based on glucose oxidase activity and peroxidase activity. The final product is a red-violet quinoneimine dye, which is used as an indicator of glucose concentration.

2.4. Induction of Pathogenesis-Related (PR) Proteins in Squash Plants

2.4.1. Experimental Design

Seeds of *C. pepo* were soaked in water for 2 days to synchronize germination and then they were sown individually in pots (15 cm diameter) with one kilogram of mixed soil that contains clay, sand, and peat moss (1:1:1). When they reached the developmental stage of four expanded leaves, they were arranged into four groups. Every 15 plants represented a treatment. The first group represented control plants, and the second group was drenched with 10 mL of bacterial culture broth on the rhizosphere in each pot (10^9 cfu mL^{-1}). Squash plants of the other two groups were covered by glass lantern, and the third group was exposed to 50 adults of unknown age or sex *B. tabaci* alone, whereas the fourth group was exposed to bacterial culture broth along with *B. tabaci* at the same rate as the second and third groups, respectively. Treatments were performed for 24, 48, 72, 96, and 120 h. All groups were kept in an insect-proof greenhouse at 25 ± 5 °C, 65 ± 5 RH, and under natural light conditions.

2.4.2. Sample Extraction for Determination of Enzymes Activity

One gram of leaf samples was collected from all plants of the four treatments groups and crushed with a pre-cooled mortar and pestle in 10 mL of 0.05 M sodium acetate buffer (pH 5.0) in the presence of 0.3 g polyvinyl pyrrolidone (PVP, also, commonly called polyvidone or povidone). Finally, samples were centrifuged at 16,000× *g* for 15 min at 4 °C. The activity of β-1,3-glucanase, chitinase, polyphenol oxidase, and peroxidase were determined using the supernatants as cruds enzymes.

2.4.3. Assay of β-1,3-Glucanase Activity

The beta-1,3-glucanase activity was colorimetrically assayed by Saikia et al. [38]. The reaction mixture consisted of 62.5 µL of 0.04% laminarin and 62.5 µL of enzyme extract. The temperature and incubation time of the reaction was 40 °C for 10 min, respectively. The reaction was stopped by adding 375 µL of dinitrosalicylic acid and heating for 5 min in a boiling water bath, and Eppendorf tubes were shaken using a vortex and their absorbance was measured at 500 nm. The enzyme activity was expressed as U g^{-1} fresh weight (quantity of enzyme that liberates one µM glucose per minute under experimental conditions).

2.4.4. Assay of Chitinase Activity

One mL of the enzyme extract was added to one mL of 1% colloidal chitin in 0.05 M citrate phosphate buffer (pH 6.6) and mixed by shaking in a test tube then kept in a water bath at 37 °C for 75 min with shaking. The reaction was then stopped by adding one mL of dinitrosalicylic acid [39] and heated for 5 min then cooled and centrifuged at 3000× *g* for 5 min to get rid of chitin before measuring O.D. at 540 nm. Chitinase activity was defined as Ug^{-1} fresh weight (µM N-Acetylglucosamine liberated per minute under experimental conditions).

2.4.5. Assay of Polyphenol Oxidase Activity

Polyphenol oxidase activity was determined according to Mayer et al. [40]. Two hundred µL of the enzyme extract was added to 1.5 mL of 0.1 M phosphate buffer (pH 7). To start the reaction, 200 µL of 0.01 M catechol in phosphate buffer (pH 7) was added and the activity was expressed as a change in absorbance at 495 nm min^{-1}.

2.4.6. Assay of Peroxidase Activity

The reaction mixture consisted of 0.5 mL of enzyme extract and 0.5 mL of 1% H_2O_2. 1.5 mL of 0.05 mL pyrogall was added to every sample separately and incubated at room temperature. The enzyme activity was expressed as the change in absorbance at 420 nm at 1 min intervals [41].

2.5. The Effect of the Bacterial Isolate on B. tabaci Population Density

To study the effect of the bacterial isolate on the attraction of *B. tabaci* adults, plants of the same germination, cultivation, pot diameter, soil weight, and developmental stage as in the enzyme assay experimental design were studied. Plants were divided into two groups in a greenhouse under the same conditions with each group consisting of five plants. The first group was the control that was drenched only by water, and the second group was drenched with 10 mL of bacterial culture broth (10^9 cfu mL^{-1}) on the rhizosphere in each pot. Whitefly adults were collected from the mother colony and released in the center of the greenhouse in the free-choice test. For each treatment, the numbers of attracted whiteflies/cm^2/plant were daily recorded and calculated as a mean of daily record. To avoid whitefly movement, the counting was done gently and in the early morning. The experiment was replicated for five successive days. The obtained data were calculated as a mean of five replicas.

2.6. The Effect of Inoculation of Squash Plants by the Bacterial Isolate on Egg-Laying and Hatchability of B. tabaci

To study the effect of bacterial isolate on the attraction of egg-laying of whitefly females and the egg hatchability, five plants inoculated with bacterial culture broth (10^9 cfu mL^{-1}) and non-inoculated plants (control) were covered by a glass lantern and infested with the natural sex ratio (2 males:3 females) of *B. tabaci* for four days to allow the oviposition. After removing the adults by shaking the plants, the laid eggs/plant were recorded and calculated as a mean of five replicas. Plants were transferred to an insect-proof greenhouse. To keep the plants isolated free from newly *B. tabaci* adults' infestation, each plant was fully covered by a glass lantern until egg emergence, and then the percentage of hatchability was recorded and calculated as a mean of five replicas.

2.7. Statistical Analysis

Randomized complete blocked design (RCBD) was used for the analysis of the obtained results (a two-way ANOVA) using SAS software [42]. The significant differences between treatments were determined according to the least significant differences (LSD) at a $p \leq 0.05$ level of probability.

3. Results

3.1. Molecular Identification of the Bacterial Isolate

The bacterial isolate used in the current study was identified based on sequencing of 16S rRNA. A search in the database to identify the bacterial isolate was performed in BLAST search at the National Center for Biotechnology Information site (http://www.ncbi.nlm.nih.gov) (accessed on 2 March 2022). The research revealed that the sequence of the investigated bacterial isolate was almost similar to several *B. velezensis* strains with a homology percentage of 99%. The PGPR isolate was identified as *B. velezensis* strain GB1 with the accession number, OM836750. A phylogenetic tree was built using the nucleotide sequence of the 16S rRNA gene of *B. velezensis* strain GB1 obtained in the current investigation and nucleotide sequences of the 16S rRNA of other Bacillus species obtained from the GenBank database (http://www.ncbi.nlm.nih.gov) (accessed on 2 March 2022). The phylogenetic tree revealed that two major clusters exist. Cluster 1 included *Escherichia coli*, whereas *B. velezensis* strain GB1obtained in this study and all *Bacillus* spp. strains provided from GenBank were clustered in Cluster 2, which was divided into two groups. The first group included *B. amyloliquifaciens*, *B. subtilis*, and *B. licheniformis* provided from GenBank, whereas the second group contained *B. velezensis* strain GB1 isolated in this investigation, and all *B. velezensis* strains collected from GenBank have a high percentage of identity that reached 95% (Figure 1).

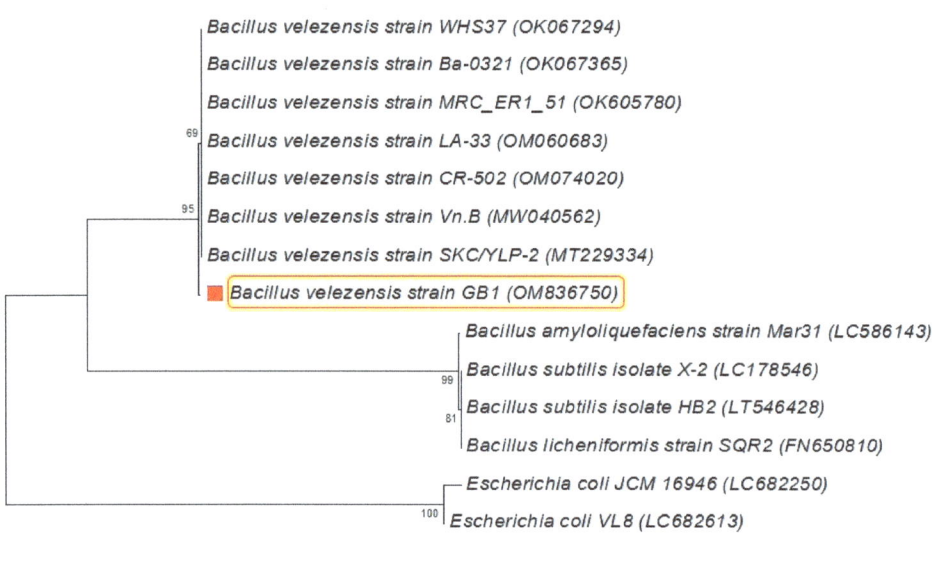

Figure 1. Phylogenetic tree of *Bacillus velezensis* strain GB1 obtained in the current study and validly described members of the genus *Bacillus* based on the nucleotide sequences of the 16S rRNA gene. The phylogenetic tree was constructed with the Neighbor-Joining method using MEGA version 5.

3.2. Production of B. velezensis Strain GB1

Batch fermentation of *B. velezensis* strain GB1 was started in a stirred tank bioreactor with an optical density ($O.D_{550nm}$) of 0.5. Biomass and glucose concentration as a carbon source of batch fermentation of *B. velezensis* strain GB1 was plotted against time, and the maximum biomass recorded was 3.8 g L^{-1} at 10.5 h (Figure 2A). Glucose concentration decreased rapidly, reaching 4.22 g L^{-1} at 5 h; however, it reached 0 g L^{-1} at 11 h. Cell mass increased exponentially over time with a constant specific growth rate (μ) of 0.09 h^{-1} (supplementary materials). One of the important factors estimated in the exponential phase of the bacterial cell growth is the yield coefficient $Y_{X/S}$, which represents the amount of obtained biomass against the amount of consumed carbon source (glucose). In this batch fermentation process, the yield coefficient recorded 0.65 g cells/g glucose (Figure 2B). Dissolved oxygen is an important factor affecting bacterial cell growth in the bioreactor, which can be controlled by agitation speed. Batch fermentation of *B. velezensis* strain GB1 was started with a low agitation speed (200 rpm) and a higher value of dissolved oxygen of 99.5%, which was reduced gradually. The dissolved oxygen percentage decreased gradually during the first two hours of the batch fermentation process to reach 21.4% at 2.15 h. The decline of dissolved oxygen percentage is an indicator of growing the bacterial culture (Figure 3). Dissolved oxygen percentage was reserved at above 20% to guarantee adequate oxygen delivery. So, agitation speed was raised gradually step by step from 200 to 600 rpm related to the growth of bacterial cells and glucose consumption from the culture broth.

3.3. Induction of Pathogenesis-Related (PR) Proteins in Squash Plants
3.3.1. β-1,3-Glucanase Activity

As shown in Table 1, squash plants treated with *B. velezensis* strain GB1 alone and *B. velezensis* + *B. tabaci* showed an increase in their levels of β-1,3-glucanase enzyme from the first day. After that, the β-1,3-glucanase level in both treatments started significantly rising to reach the maximum value of 40.34 U g^{-1} fresh weight on the fourth day (2.98 folds over control) and 42.21 U g^{-1} fresh weight on the fifth day (2.87 folds over control), respectively. Although that *B. tabaci*-infested squash plants showed a significant enhancement in the activity of β-1,3-glucanase from the second day of infestation compared with control and fluctuated high and low until the fifth day, it remained significantly less than what was recorded by exposure to *B. velezensis* alone or with *B. tabaci* and *B. velezensis*.

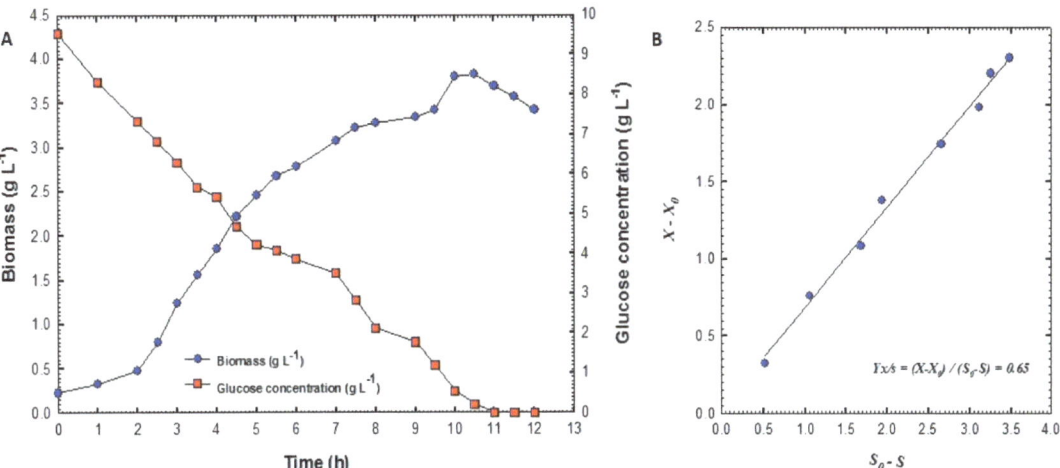

Figure 2. (**A**) Biomass and glucose concentration of fermentation broth as a function of time for batch fermentation of *Bacillus velezensis* strain GB1, and (**B**) yield coefficient for growth of *Bacillus velezensis* strain GB1 on glucose. X_0 represents the cell mass in the broth at initial time t_0 (g L^{-1}); X, the cell mass concentration in the broth at time t (g L^{-1}); S_0, glucose concentration in the broth at initial time t_0 (g L^{-1}); and S, glucose concentration in the broth at time t (g L^{-1}).

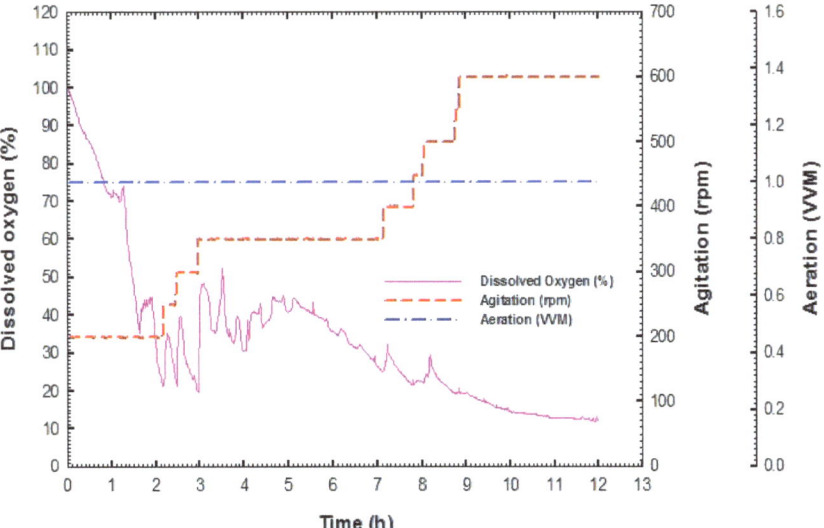

Figure 3. Dissolved oxygen, agitation, and aeration as a function of time during batch fermentation of *Bacillus velezensis* strain GB1.

Table 1. The time-course of β-1,3-glucanase activity in squash plants exposed to *Bemisia tabaci* and *Bacillus velezensis*, and both agents compared with control.

Treatments	β-1,3-Glucanase Activity (U g^{-1} Fresh Weight)				
	24 h	48 h	72 h	96 h	120 h
Control	* 15.11 ± 0.99 ^{jk,**}	14.83 ± 0.76 ^{k}	16.34 ± 1.68 ^{j}	13.52 ± 1.42 ^{k}	14.70 ± 0.82 ^{k}
B. tabaci	15.31 ± 0.81 ^{jk}	23.69 ± 0.51 ^{g}	20.29 ± 0.43 ^{h}	18.81 ± 0.75 ^{i}	25.22 ± 0.51 ^{f}
B. velezensis	24.73 ± 0.67 ^{f}	33.66 ± 0.63 ^{d}	35.79 ± 0.71 ^{c}	40.34 ± 0.55 ^{b}	40.20 ± 0.51 ^{b}
B. velezensis + B. tabaci	30.76 ± 0.87 ^{e}	29.71 ± 1.4 ^{e}	35.86 ± 0.31 ^{c}	40.17 ± 0.19 ^{b}	42.21 ± 0.47 ^{a}

* Means in each column followed by the same letter do not differ significantly ($p \leq 0.05$); ** significant letters.

3.3.2. Chitinase Activity

As illustrated in Table 2, the accumulation of chitinase started to increase with time one day after exposure to the tested agents with the absence of a considerable change in the case of *B. tabaci*. Treating squash plants with *B. velezensis* strain GB1 induced a significant accumulation of chitinase to reach the maximum level of 0.0497 U g^{-1} fresh weight on the fifth day (9 fold more than control), which is the most significant level of activity among all treatments. Also, the presence of bacteria in combination with the insect resulted in the enhancement of the chitinase level from the first day, which then gradually increased with time to reach the maximum increase of 0.0486 U g^{-1} fresh weight on the fourth day (8.5 fold more than control). The chitinase accumulation in the squash leaves during the first three days of exposure to *B. tabaci* was almost equal and without any significant difference to its level in the control, which was then slightly elevated on the fourth and fifth days (1.2 and 1.4 fold increase compared to control, respectively).

Table 2. The time-course of chitinase activity in squash plants exposed to *Bemisia tabaci* and *Bacillus velezensis*, and both agents compared with control.

Treatments	Chitinase Activity (U g^{-1} Fresh Weight)				
	24 h	48 h	72 h	96 h	120 h
Control	* 0.0054 ± 0.00015 ^{j,**}	0.0055 ± 0.00025 ^{j}	0.0053 ± 0.0002 ^{j}	0.0057 ± 0.00011 ^{j}	0.0055 ± 0.0001 ^{j}
B. tabaci	0.0057 ± 0.00005 ^{j}	0.0060 ± 0.00011 ^{j}	0.0066 ± 0.0001 ^{ij}	0.0070 ± 0.0001 ^{i}	0.0078 ± 0.00025 ^{h}
B. velezensis	0.0142 ± 0.0003 ^{g}	0.0379 ± 0.00052 ^{e}	0.0426 ± 0.0010 ^{d}	0.0474 ± 0.0007 ^{c}	0.0497 ± 0.0006 ^{a}
B. velezensis + B. tabaci	0.01843 ± 0.0006 ^{f}	0.0378 ± 0.0004 ^{e}	0.0383 ± 0.0005 ^{e}	0.0486 ± 0.00005 ^{b}	0.0382 ± 0.0005 ^{e}

* Means in each column followed by the same letter do not differ significantly ($p \leq 0.05$); ** significant letters.

3.3.3. Polyphenol Oxidase Activity

As shown in Table 3, inoculation with *B. velezensis* strain GB1 recorded the most significant treatment in increasing the level of polyphenol oxidase in squash plants where it reached the highest level of 28.82 U g^{-1} fresh weight on the second day (4.6 fold more than control). Subsequently, the enzyme activity retreated gradually with time. The treatment with *B. velezensis* + *B. tabaci* and infestation with *B. tabaci* alone recorded the maximum levels of polyphenol oxidase enzyme of 24.15 and 19.20 U g^{-1} fresh weight on the third day of the treatments, respectively, then gradually retracted at 96 and 120 h post-treatment.

Table 3. The time-course of polyphenol oxidase activity in squash plants exposed to *Bemisia tabaci* and *Bacillus velezensis*, and both agents compared with control.

Treatments	Polyphenol Oxidase Activity (U g^{-1} Fresh Weight)				
	24 h	48 h	72 h	96 h	120 h
Control	* 4.33 ± 0.02 ^{n,**}	6.25 ± 0.09 ^{l}	3.80 ± 0.23 ^{o}	4.71 ± 0.16 ^{n}	5.43 ± 0.4 ^{m}
B. tabaci	9.73 ± 0.55 ^{k}	14.17 ± 0.045 ^{j}	19.20 ± 0.20 ^{h}	18.62 ± 0.36 ^{i}	14.15 ± 0.09 ^{j}
B. velezensis	25.48 ± 0.22 ^{c}	28.82 ± 0.3 ^{a}	27.82 ± 0.43 ^{b}	25.87 ± 0.08 ^{c}	23.82 ± 0.29 ^{d}
B. velezensis + B. tabaci	22.16 ± 0.50 ^{f}	23.25 ± 0.8 ^{e}	24.15 ± 0.19 ^{d}	19.12 ± 0.14 ^{h}	20.34 ± 0.17 ^{g}

* Means in each column followed by the same letter do not differ significantly ($p \leq 0.05$); ** significant letters.

3.3.4. Peroxidase Activity

As displayed in Table 4, peroxidase activity was affected by different treatments. It was observed that the maximum rate of activity was significantly exhibited on the fifth day of inoculation with *B. velezensis* strain GB1 to reach 1883.67 U g^{-1} fresh weight, which is equivalent to 1.87 fold greater than the control plant at the same time. The combined effect of *B. velezensis* strain GB1 and the whitefly on the peroxidase activity was evident from the first day of exposure. The peroxidase activity gradually increased with time to reach the maximum value on the fourth day to reach 1801.33 U g^{-1} fresh weight (1.82 fold more than control). Feeding of *B. tabaci* induced the accumulation of peroxidase from the beginning and then progressively increased with time to reach the ultimate of 1375.67 U g^{-1} fresh weight (1.4 folds more than control) on the fourth day.

Table 4. The time-course of peroxidase activity in squash plants exposed to *Bemisia tabaci* and *Bacillus velezensis*, and both agents compared with control.

Treatments	Peroxidase Activity (U g^{-1} Fresh Weight)				
	24 h	48 h	72 h	96 h	120 h
Control	* 932.67 ± 7.63 [m,**]	1004.66 ± 7.37 [k]	987.00 ± 14.52 [l]	985.66 ± 10.07 [l]	1005.33 ± 9.07 [k]
B. tabaci	1187.33 ± 9.71 [j]	1235.00 ± 4.00 [i]	1351.33 ± 22.01 [h]	1375.67 ± 5.03 [gh]	1363.67 ± 5.13 [h]
B. velezensis	1535.33 ± 15.63 [f]	1542.67 ± 11.59 [f]	1785.00 ± 15.1 [cd]	1838.66 ± 6.51 [b]	1883.67 ± 12.58 [a]
B. velezensis + B. tabaci	1387.67 ± 7.02 [g]	1390.33 ± 8.02 [g]	1782.33 ± 13.01 [d]	1801.33 ± 9.50 [c]	1764.00 ± 14.11 [c]

* Means in each column followed by the same letter do not differ significantly ($p \leq 0.05$); ** significant letters.

3.4. The Effect of the Inoculation of Squash Plants with B. velezensis Strain GB1 on B. tabaci Population Density

As shown in Figure 4A, the mean number of *B. tabaci* adults/cm^2 in the control did not show any significant difference throughout the experiment period from the first day (3.63 adults/cm^2) to the end of the fifth day (3.77 adults/cm^2). On the other hand, the mean number of *B. tabaci* in the case of *B. velezensis*-inoculated squash plants decreased significantly from the first day compared to that recorded on any day of the control. The mean number of *B. tabaci* recorded 2.56 adults/cm^2 on the first day then it decreased over time with a significant decrease to reach the lowest significant level of 0.95 adult/cm^2 on the fifth day. Figure 4B shows the rate of the reduction of *B. tabaci* attraction to *B. velezensis*-inoculated squash plants with time. The reduction percentage of attraction on the first day as a consequence of inoculation with strain GB1was 29.5%, which increased gradually to reach 74.03% on the fifth day. From the obtained results, it seems clear that the inoculation of squash plants with *B. velezensis* strain GB1 plays an important defensive role against whiteflies. It was observed that the inoculation of squash plants with strain GB1 reduced the mean number of whitefly adults/cm^2 by about 49.3% throughout the experiment period.

3.5. The Effect of the Inoculation of Squash Plants with B. velezensis Strain GB1 on Egg-Laying and Hatchability of B. tabaci

The obtained results from statistical analysis referred to a significant negative correlation between the inoculation of squash plants with strain GB1 and the oviposition rate. As shown in Figure 5A, the number of laid eggs per female per day was decreased by the inoculation of squash plants with strain GB1. Whiteflies fed on squash plants inoculated with *B. velezensis*, and laid 2.28 eggs/female/day, which was significantly lower by about 43.98% than the mean number of laid eggs by whiteflies fed on non-inoculated plants (4.07 eggs/female/day). The percentages of the egg hatchability take the same trend of oviposition. As shown in Figure 5B, the percentage of the egg hatchability in normal plants was 91.35%, which was significantly higher than its value of 86.11% in the inoculated plants, which was considered a 5.7% decline in hatchability percentage.

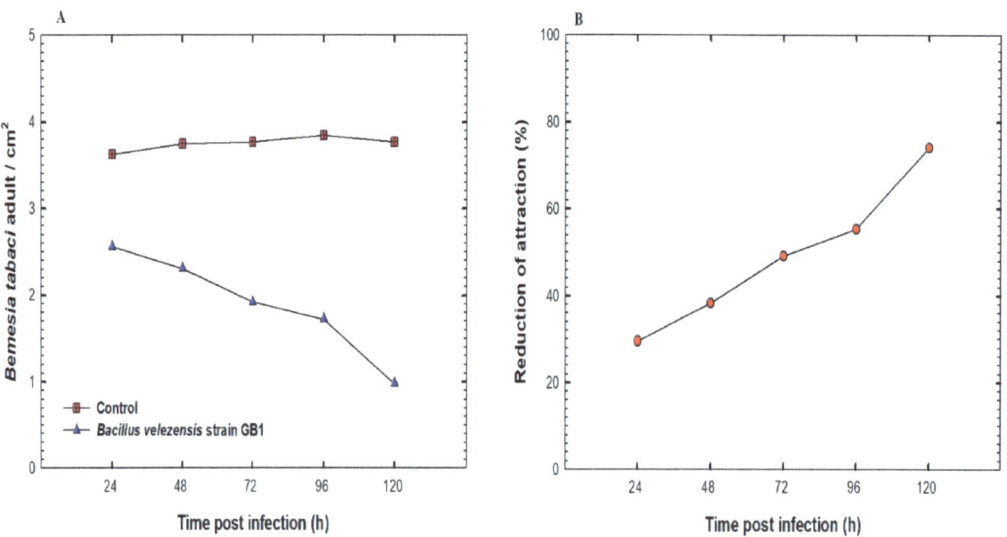

Figure 4. The effect of squash plants' inoculation with *Bacillus velezensis* Stain GB1 on (**A**) the population density and (**B**) the attraction reduction percentage of *Bemisia tabaci*.

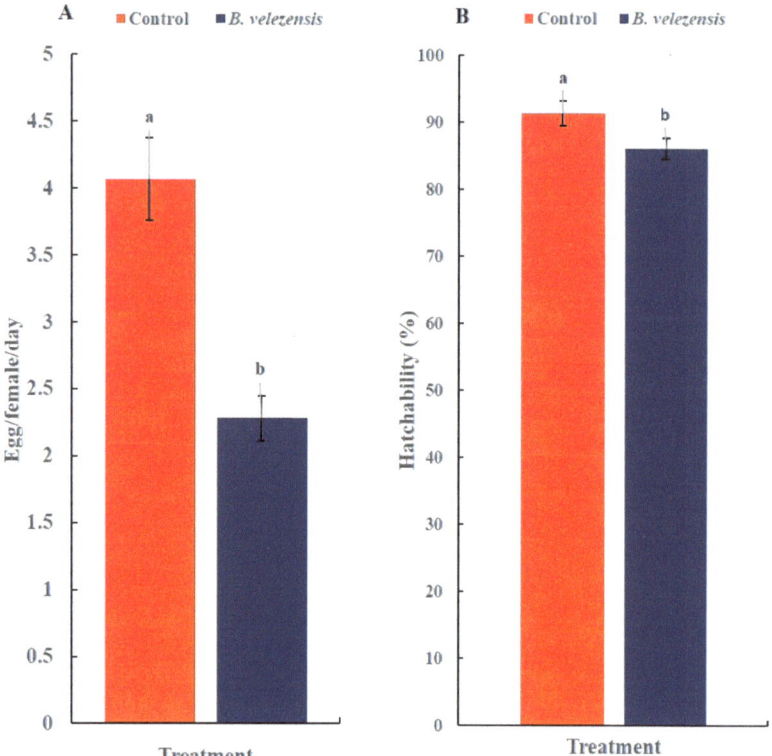

Figure 5. The effect of squash plants' inoculation with *Bacillus velezensis* on (**A**) the daily mean number and (**B**) hatchability of laid eggs by *Bemisia tabaci* females.

4. Discussion

Recently, integrated pest management approaches play an important role in reducing the role of chemical pesticides in the insect management approach, based on natural resources and behaviors such as host resistance and biological control [43]. Effectual control of pests with little or no natural environmental hazard is accomplished by biological control using microorganisms as a good alternative approach to chemical control [44]. Nelson [45] reported a prospective strategy for effective biological control that was performed using plant growth-promoting rhizobacteria (PGPR). So, the main target of the present study is maximizing the production of *B. velezensis* strain GB1 as a biocontrol agent using a batch fermentation process in the stirred tank bioreactor for biological control of whitefly *B. tabaci* on squash plants.

Cultivation in the stirred tank bioreactor compared to a shake flask provides the bacterial culture with the optimum conditions of temperature, pH, agitation, and aeration that are required for maximizing the growth of the bacterial cells and secreted secondary metabolites. Biomass of *B. velezensis* strain GB1 increased exponentially during the exponential phase with a constant specific growth rate of 0.09 h^{-1} to reach a high value of 3.8 g L^{-1} at 10.5 h with a production rate equal to 0.4 g L^{-1} h^{-1}. Matar et al. [35] used the batch fermentation process for scaling-up production of biomass of a biocontrol agent, *B. subtilis* isolate G-GANA7. The maximum biomass obtained by them reached 3.2 g L^{-1} at 11 h. The batch fermentation process of *B. velezensis* strain GB1 achieved a high yield of biomass, which recorded 0.65 g cell/g glucose. The same yield coefficient of biomass was achieved by Abdel-Gayed et al. [46] through a batch fermentation process of the biocontrol agent, *B. subtilis* isolate B4. In the exponential phase of our bacterial growth curve, the bacterial cells grow rapidly and consume a lot amount of oxygen and carbon source necessitated for growing with a consumption rate of glucose equivalent to 0.8 g L^{-1} h^{-1}. Our results revealed that the dissolved oxygen decreased gradually during the first two hours of the culture growth. Afterward, with the beginning of the exponential phase, the dissolved oxygen decreased rapidly so the agitation speed was raised to reserve it at above 20%.

Induced systemic resistance in plants against bacterial and fungal pathogens by *B. velezensis* has been established. Nevertheless, their activities against insect attack, and fundamental cellular and molecular defense mechanisms have not been illuminated up until now [47]. Molecular and physiological improvements in plants are activated by PGPR, mediating boosted plant defense against pathogens and insect pests by inducing systemic resistance (ISR) [47–50]. The present study revealed that the inoculation of squash plants by *B. velezensis* strain GB1 induced systemic resistance against *B. tabaci*. Rashid et al. [47] documented that *B. velezensis* YC7010 induced systemic resistance against the green peach aphid (GPA), *Myzus persicae*. Additionally, *B. velezensis* YC7010 was utilized as an inducer to systemic resistance of rice against brown planthopper (BPH; *Nilaparvata lugens* Stål), which is one of the most serious insect pests that reduces rice yield remarkably in many rice-growing areas [33]. In the current study, β-1,3-glucanase activity, chitinase activity, polyphenol oxidase activity, and peroxidase activity have been evaluated in squash plants treated with *B. tabaci* alone, *B. velezensis* strain GB1 alone, and *B. tabaci* + *B. velezensis* strain GB1 compared to control treatment. A positive correlation has been recorded between the inoculation of *B. velezensis* strain GB1, *B. tabaci* feeding, and the activity of all tested enzymes as follows:

In our study, PGPR *B. velezensis* strain GB1 has enhanced β-1,3-glucanase activity in squash plants against whitefly *B. tabaci* with a significant increase. Moreover, the present data indicate the presence of an effect of the whitefly feeding on the enzymatic activity of the squash plants as a natural and defensive reaction of the plant. These results are closed to the obtained results by Jimenez et al. Mayer et al. and Inbar et al. [51–53] showed that infestation of whiteflies resulted in the aggregation of pathogenesis-related proteins in tomato and other plants, which were thought to have a defensive role against insect pests. Gene-encoding pathogenesis-related proteins were induced locally in squash following silverleaf whitefly feeding [54]. Many studies confirmed that *B. tabaci* feeding boosted the activity of pathogenesis-related proteins such as β-1,3-glucanase, chitinase, and peroxidase in cassava, tomato, black gram, and tobacco (*Vigna mungo*) compared to non-infested plants [55–59].

In the present study, *B. velezensis* strain GB1 alone induced chitinase activity to reach its maximum level on the fifth day followed by treatment with strain GB1 in the presence of *B. tabaci* infection. The mode of action of the chitinase enzyme is based on the degradation of chitin, a critical element of insect cells [60]. It has the potential to harm insects by destroying chitin-based structures such as the peritrophic membrane that provides a physical barrier to ingested pathogens and other substances that could be considered a danger to the insect [61]. The poplar chitinases were used as a method to control and inhibit the development of the Colorado potato beetle and pest population, which hinted at its involvement in the tomato plant defense [62].

In the current study, the application of *B. velezensis* strain GB1 has induced polyphenol oxidase activity as a defense enzyme in response to whitefly *B. tabaci*. *B. velezensis* XT1 increased polyphenol oxidase activity by 395%, indicating an enhancement of olive trees' resistance to *Verticillium dahlia* [63]. The inoculation of cotton roots with *B. velezensis* triggered induced systemic resistance (ISR) against *V. dahlia* and caused the activation of the antioxidant enzymes such as phenylalanine ammonia-lyase, polyphenol oxidase, peroxidase, and phenol contents [64]. Polyphenol oxidases exist in many plants [65]. They are metalloenzymes that contain a type-3 copper center and function as defense enzymes [66]. Polyphenol oxidase oxidizes phenolics to create highly reactive o-quinones that increase the anti-insect activity of phenolics. Quinones covalently link to proteins, reducing their permeability as nutritional resources [67,68]. Anti-nutritive action of polyphenol oxidase against insects is based on the formation of reactive oxygen species (e.g., superoxide radical and H_2O_2) that harm essential nutrients or fundamental elements for insects such as proteins, lipids, and nucleic acids [69–71]. The present study is in agreement with the studies on the Colorado potato beetle as a considerable positive correlation between polyphenol oxidase levels and larval mortality [72]. Furthermore, Bhonwong et al. [73] reported that overexpression of a polyphenol oxidase in tomato led to a reduction in the growth average and nutritional clue of cotton bollworm (*Helicoverpa armigera*) and beet armyworm (*Spodoptera exigua*), which also improved resistance against *Spodoptera litura* and *Malacosoma disstria* in tomato and poplar, respectively [74,75].

In our study, *B. velezensis* strain GB1 induced peroxidase accumulation in squash plants infected with whitefly *B. tabaci*. Peroxidase has been associated with a multitude of physiological processes, which could help plants stand against herbivore and pathogen attacks including polysaccharide crosslinking, oxidation of hydroxyl cinnamyl alcohol into free radical intermediates, phenol oxidation, hypersensitive response, crosslinking of extension monomers and lignification, production and polymerization of phenolics, negative effects on food digestibility, and protein availability to sucking pests [76,77]. Infestation with leaf miners resulted in an accumulation of peroxidase isoforms in groundnut plants [78]. Peroxidase-oxidized phenolics within the herbivore stomach and induced the production of semi-quinone radicals in the midguts of larvae, resulting in a lower growth rate of larvae that consumed poplars with induced peroxidase activity [79]. Dowd and Lagrimini [80] noticed that a high level of peroxidase activity in the infested plants with *Trialeurodes vaporariorum* whiteflies resulted in a reduction of the population density of whiteflies per plant. These results were closed to the obtained results in the present study. After treatment with *B. velezensis* SDTB022, the activities of defense enzymes in tobacco, such as polyphenol oxidase, peroxidase, and phenylalanine ammonia-lyase, significantly increased and, consequently, plant yield and disease defense responses in the field increased [81].

The present study showed that the inoculation of squash plants with *B. velezensis* GB1 significantly decreased the mean number of attracted *B. tabaci* and the percentage of the egg hatchability. The mixtures of the *B. velezensis* strain were responsible for emitting higher amounts of plant volatiles in cotton plants following *Heliothis virescens* larvae infestation [82].

The combination of *B. velezensis* strains that was referred to as Blend-8 and Blend-9 induced cotton resistance, and reduced the growth and development of *S. exigua* via the increased level of gossypol [24]. Systemic resistance in Arabidopsis seedlings against *M. persicae* was induced by inoculation with *B. velezensis*, which significantly reduced the settling, feeding, and reproduction of *M. persicae* on Arabidopsis leaves [47]. *Ostrinia nubilalis* laid significantly fewer eggs on maize plants treated with the different PGPR strains including *B. velezensis* compared to untreated plants, which can change maize plant volatiles with imperative ramifications for plant–insect interactions [83]. A mixture of lipopeptides (iturins, surfactins, and fengycins) was produced by *B. velezensis* B64a and *B. velezensis* B15, which were the best bioagent against *Aedes aegypti* [84]. *B. subtilis*-inoculated cotton plants showed raised levels of polyphenol oxidase, peroxidase, and chitinase, which had a substantial influence on reducing the aphid population under greenhouse conditions up to 14 days following treatment [85]. Hanafi et al. [86] found that inoculating tomato plants with the *B. subtilis* strain resulted in considerably reduced survival of *B. tabaci*'s nymphs and pupae, implying that inoculation with *B. subtilis* confers some type of resistance or avoidance behavior to plants, resulting in less *B. tabaci* propagation on the *B. subtilis*-inoculated plants. Valenzuela-Soto et al. [87] demonstrated that tomato plants treated with *B. subtilis* not only enhanced plant growth but also resulted in the establishment of an ISR through induction of several PR protein genes, resulting in the significant reduction in 4th instar nymphs, pupae, empty pupal cases (corresponding emerging adults), and emerged adults, with *B. tabaci* development favored over untreated controls.

5. Conclusions

The batch fermentation process of *B. velezensis* strain GB1 achieved in the stirred tank bioreactor was sufficient in maximizing the production of secondary metabolites and culture biomass, which reached a maximum value of 3.8 g L^{-1} with a yield coefficient of 0.65 g cells/g glucose. *B. velezensis* strain GB1 induced squash plants to enhance their levels of β-1,3-glucanase, chitinase, polyphenol oxidase, and peroxidase enzymes. Additionally, *B. velezensis* strain GB1 was effective in reducing the mean number of the attracted *B. tabaci* on squash plants with a significant decrease. The reduction percentage of *B. tabaci* attraction reached 74.03% on the fifth day. Squash plants inoculated with *B. velezensis* strain GB1 recorded a lower mean number of *B. tabaci* eggs/female/day compared to control with a decreasing percentage of 43.98%. The percentage of egg hatchability on squash plants inoculated with *B. velezensis* strain GB1 declined by 5.7%. Finally, *B. velezensis* strain GB1 could be considered one of the important biocontrol agents that plays a pivotal role in plant protection against whitefly *B. Tabaci* and a promising tool for integrated pest management (IPM) programs.

Author Contributions: Planning and designing the research, A.S. and G.A.-Z.; methodology, A.S. and G.A.-Z.; data analysis and software, A.S. and G.A.-Z.; writing—original draft preparation, A.S. and G.A.-Z.; writing—review and editing, A.S., G.A.-Z. and S.M. All authors have read and agreed to the published version of the manuscript.

Funding: This research received no external funding.

Informed Consent Statement: Not applicable.

Conflicts of Interest: The authors declare no conflict of interest.

References

1. Oliveira, M.R.V.; Henneberry, T.J.; Anderson, P. History, current status and collaborative research projects for *Bemisia tabaci*. *Crop Prot.* **2001**, *20*, 709–723. [CrossRef]
2. Hoelmer, K.A.; Osborne, L.S.; Yokomi, R.K. Foliage disorders in Florida associated with feeding by sweet potato whitefly, *Bemisia tabaci*. *Fla. Entomol.* **1991**, *74*, 162–166. [CrossRef]
3. Summers, C.G.; Estrada, D. Chlorotic streak of bell pepper: A new toxicogenic disorder induced by feeding of the Silverleaf whitefly, *Bemisia argentifolii*. *Plant Dis.* **1996**, *80*, 822. [CrossRef]

4. McAuslane, H.J.; Cheng, J.; Carle, R.B.; Schmalstig, J. Influence of *Bemisia argentifolii* (Homoptera: Aleyrodidae) infestation and squash silver leaf disorder on zucchini seedling growth. *J. Econ. Entomol.* **2004**, *97*, 1096–1105. [CrossRef]
5. Brown, J.K.; Czosnek, H. Whitefly transmission of plant viruses. *Adv. Bot. Res.* **2002**, *36*, 65–92.
6. Jones, D.R. Plant viruses transmitted by whiteflies. *Eur. J. Plant Pathol.* **2003**, *109*, 195–219. [CrossRef]
7. Ng, J.C.; Tian, T.; Falk, B.W. Quantitative parameters determining whitefly (*Bemisia tabaci*) transmission of lettuce infectious yellows virus and an engineered defective RNA. *J. Gen. Virol.* **2004**, *86*, 2697–2707. [CrossRef]
8. Al-Momany, A.; Al-Antary, T.M. *Pests of Garden and Home*, 2nd ed.; Publications of University of Jordan: Amman, Jordan, 2008; p. 518.
9. Al Arabiat, O.W.; Araj, S.A.; Alananbeh, K.M.; Al-Antary, T.M. Effect of three *Bacillus* spp. on tobacco whitefly *Bemisia tabaci* (Gennadius) (Homoptera: Aleyrodidae). *Fresen. Environ. Bull.* **2018**, *27*, 3706–3712.
10. Sani, I.; Ismail, S.I.; Abdullah, S.; Jalinas, J.; Jamian, S.; Saad, N. A review of the biology and control of whitefly, *Bemisia tabaci* (Hemiptera: Aleyrodidae), with special reference to biological control using entomopathogenic fungi. *Insects* **2020**, *11*, 619. [CrossRef]
11. Bellotti, A.C.; Arias, B. Host plant resistance to whiteflies with emphasis on cassava as a case study. *Crop Prot.* **2001**, *20*, 813–823. [CrossRef]
12. Kant, M.R.; Jonckheere, W.; Knegt, B.; Lemos, F.; Liu, J.; Schimmel, B.C.J.; Villarroel, C.A.; Ataide, L.M.S.; Dermauw, W.; Glas, J.J.; et al. Mechanisms and ecological consequences of plant defense induction and suppression in herbivore communities. *Ann. Bot.* **2015**, *115*, 1015–1051. [CrossRef] [PubMed]
13. Ali, S.; Ganai, B.A.; Kamili, A.N.; Bhat, A.A.; Mir, Z.A.; Bhat, J.A.; Tyagi, A.; Islam, S.T.; Mushtaq, M.; Yadav, P.; et al. Pathogenesis-related proteins and peptides as promising tools for engineering plants with multiple stress tolerance. *Microb. Res.* **2018**, *212–213*, 29–37. [CrossRef] [PubMed]
14. Bashan, Y.; De Bashan, L. Plant growth-promoting. *Encycl. Soils Environ.* **2005**, *1*, 103–115.
15. Al Arabiat, O.W.; Araj, S.E.A.; Alananbeh, K.M.; Al-Antary, T.M. Efficacy of three *Bacillus* spp. on development of tobacco whitefly *Bemisia tabaci* (Gennadius) (Homoptera: Aleyrodidae). *Fresen. Environ. Bull.* **2018**, *27*, 4965–4972.
16. Van Loon, L.C. Plant responses to plant growth-promoting rhizobacteria. *Eur. J. Plant Pathol.* **2007**, *119*, 243–254. [CrossRef]
17. Chandrasekaran, M.; Chun, S.C. Expression of PR-protein genes and induction of defense-related enzymes by *Bacillus subtilis* CBR05 in tomato (*Solanum lycopersicum*) plants challenged with *Erwinia carotovora* subsp. *carotovora*. *Biosci. Biotechnol. Biochem.* **2016**, *80*, 2277–2283. [CrossRef]
18. Van Hulten, M.; Pelser, M.; Van Loon, L.C.; Pieterse, C.M.; Ton, J. Costs and benefits of priming for defense in Arabidopsis. *Proc. Nat. Acad. Sci. USA* **2006**, *103*, 5602–5607. [CrossRef]
19. Chung, E.J.; Hossain, M.T.; Khan, A.; Kim, K.H.; Jeon, C.O.; Chung, Y.R. *Bacillus oryzicola* sp. nov., an endophytic bacterium isolated from the roots of rice with antimicrobial, plant growth promoting, and systemic resistance inducing activities in rice. *Plant Pathol. J.* **2015**, *31*, 152. [CrossRef]
20. Hossain, M.T.; Khan, A.; Chung, E.J.; Rashid, H.; Chung, Y.R. Biological control of rice bakanae by an endophytic *Bacillus oryzicola* YC7007. *Plant Pathol. J.* **2016**, *32*, 228–241. [CrossRef]
21. Lugtenberg, B.; Kamilova, F. Plant-growth-promoting rhizobacteria. *Annu. Rev. Microbiol.* **2009**, *63*, 541–556. [CrossRef]
22. Pieterse, C.M.; Van der Does, D.; Zamioudis, C.; Leon-Reyes, A.; Van Wees, S.C. Hormonal modulation of plant immunity. *Annu. Rev. Cell Dev. Biol.* **2012**, *28*, 489–521. [CrossRef] [PubMed]
23. DeOliveira Araujo, E. Rhizobacteria in the control of pest insects in agriculture. *Afr. J. Plant Sci.* **2015**, *9*, 368–373.
24. Zebelo, S.; Song, Y.; Kloepper, J.W.; Fadamiro, H. Rhizobacteria activates (C)-D-cadinene synthase genes and induces systemic resistance in cotton against beet armyworm (*Spodoptera exigua*). *Plant Cell Environ.* **2016**, *9*, 935–943. [CrossRef] [PubMed]
25. Zehnder, G.; Kloepper, J.; Tuzun, S.; Yao, C.; Wei, G.; Chambliss, O.; Shelby, R. Insect feeding on cucumber mediated by rhizobacteria- induced plant resistance. *Entomol. Exp. Appl.* **1997**, *83*, 81–85. [CrossRef]
26. Ruiz-García, C.; Bejar, V.; Martinez-Checa, F.; Llamas, I.; Quesada, E. *Bacillus velezensis* sp. nov., a surfactant-producing bacterium isolated from the river Velez in Malaga, southern Spain. *Int. J. Syst. Evol. Microbiol.* **2005**, *55*, 191–195. [CrossRef]
27. Cao, Y.; Pi, H.; Chandrangsu, P.; Li, Y.; Wang, Y.; Zhou, H.; Xiong, H.; Helmann, J.D.; Cai, Y. Antagonism of two plant-growth promoting *Bacillus velezensis* isolates against *Ralstonia solanacearum* and *Fusarium oxysporum*. *Sci. Rep.* **2018**, *8*, 4360. [CrossRef]
28. Chen, L.; Shi, H.; Heng, J.; Wang, D.; Bian, K. Antimicrobial, plant growth-promoting and genomic properties of the peanut endophyte *Bacillus velezensis* LDO2. *Microbiol. Res.* **2019**, *218*, 41–48. [CrossRef]
29. Meng, Q.; Jiang, H.; Hao, J.J. Effects of *Bacillus velezensis* strain BAC03 in promoting plant growth. *Biol. Control* **2016**, *98*, 18–26. [CrossRef]
30. Kim, S.Y.; Song, H.; Sang, M.K.; Weon, H.Y.; Song, J. The complete genome sequence of *Bacillus velezensis* strain GH1-13 reveals agriculturally beneficial properties and a unique plasmid. *J. Biotechnol.* **2017**, *259*, 221–227. [CrossRef]
31. Fan, B.; Wang, C.; Song, X.; Ding, X.; Wu, L.; Wu, H.; Gao, X.; Borriss, R. *Bacillus velezensis* FZB42 in 2018: The gram-positive model strain for plant growth promotion and biocontrol. *Front. Microbiol.* **2018**, *9*, 2491. [CrossRef]
32. Rabbee, M.F.; Ali, M.; Choi, J.; Hwang, B.S.; Jeong, S.C.; Baek, K.H. *Bacillus velezensis*: A valuable member of bioactive molecules within plant microbiomes. *Molecules* **2019**, *24*, 1046. [CrossRef] [PubMed]

33. Rashid, M.H.; Kim, H.J.; Yeom, S.I.; Yu, H.A.; Manir, M.M.; Moon, S.S.; Kang, Y.J.; Chung, Y.R. *Bacillus velezensis* YC7010 enhances plant defenses against brown planthopper through transcriptomic and metabolic changes in rice. *Front. Plant Sci.* **2018**, *9*, 1904. [CrossRef] [PubMed]
34. Istock, C.A.; Ferguson, N.; Istock, N.L.; Duncan, K.E. Geographical diversity of genomic lineages in *Bacillus subtilis* (Ehrenberg) Cohn sensulato. *Org. Divers. Evol.* **2001**, *1*, 179–191. [CrossRef]
35. Matar, S.M.; El-Kazzaz, S.A.; Wagih, E.E.; El-Diwany, A.I.; Hafez, E.E.; Moustafa, H.E.; Abo-Zaid, G.A.; Serour, E.A. Molecular characterization and batch fermentation of *Bacillus subtilis* as biocontrol agent II. *Biotechnology* **2009**, *8*, 35–43. [CrossRef]
36. Asaka, O.; Shoda, M. Biocontrol of *Rhizoctonia solani* damping-off of tomato with *Bacillus subtilis* RB14. *Appl. Environ. Microb.* **1996**, *62*, 4081–4408. [CrossRef]
37. Van Dam-Mieras, M.C.E.; Jeu, W.H.; Vries, J.; Currell, B.R.; James, J.W.; Leach, C.K.; Patmore, R.A. *Techniques Used in Bioproduct Analysis*; Butterworth-Heinemann Ltd.: Oxford, UK, 1992.
38. Saikia, R.; Singh, B.P.; Kumar, R.; Arora, D.K. Detection of pathogenesis-related proteins-chitinase and β-1,3-glucanase in induced chickpea. *Curr. Sci.* **2005**, *89*, 659–663.
39. Monreal, J.; Reese, E.T. The chitinase of *Serratia marcescens*. *Can. J. Microb.* **1969**, *15*, 689–696. [CrossRef]
40. Mayer, A.M.; Harel, E.; Shaul, R.B. Assay of catechol oxidase a critical comparison of methods. *Phytochemistry* **1965**, *5*, 783–789. [CrossRef]
41. Hammerschmidt, R.; Nuckles, E.M.; Kuc, J. Association of enhanced peroxidase activity with induced systemic resistance of cucumber to *Colletotrichum lagenarium*. *Physiol. Plant Pathol.* **1982**, *20*, 73–76. [CrossRef]
42. SAS (Statistical Analysis System). *SAS/STAT User's Guide*, 8.6th ed.; SAS institute Inc.: Cary, NC, USA, 2002.
43. Xu, H.-X.; Yang, Y.-J.; Lu, Y.-H.; Zheng, X.-S.; Tian, J.-C.; Lai, F.-X.; Fu, Q.; Lu, Z.-X. Sustainable management of rice insect pests by non-chemical-insecticide technologies in China. *Rice Sci.* **2017**, *24*, 61–72.
44. Li, M.; Li, S.; Xu, A.; Lin, H.; Chen, D.; Wang, H. Selection of *Beauveria* isolates pathogenic to adults of *Nilaparvata lugens*. *J. Insect Sci.* **2014**, *14*, 32. [CrossRef] [PubMed]
45. Nelson, L.M. Plant growth promoting rhizobacteria (PGPR): Prospects for new inoculants. *Crop Manag.* **2004**, *10*, 301–305. [CrossRef]
46. Abdel-Gayed, M.A.; Abo-Zaid, G.A.; Matar, S.M.; Hafez, E.E. Fermentation, formulation and evaluation of PGPR *Bacillus subtilis* isolate as a bioagent for reducing occurrence of peanut soil-borne diseases. *J. Integr. Agric.* **2019**, *18*, 2080–2092.
47. Rashid, M.H.; Khan, A.; Hossain, M.T.; Chung, Y.R. Induction of systemic resistance against aphids by endophytic *Bacillus velezensis* YC7010 via expressing PHYTOALEXIN DEFICIENT4 in Arabidopsis. *Front. Plant Sci.* **2017**, *8*, 211. [CrossRef]
48. Van Oosten, V.R.; Bodenhausen, N.; Reymond, P.; Van Pelt, J.A.; Van Loon, L.C.; Dicke, M.; Pieterse, C.M. Differential effectiveness of microbially induced resistance against herbivorous insects in *Arabidopsis*. *Mol. Plant Microbe Interact.* **2008**, *21*, 919–930. [CrossRef]
49. De Vleesschauwer, D.; Höfte, M. Rhizobacteria-induced systemic resistance. *Adv. Bot. Res.* **2009**, *51*, 223–281.
50. Pineda, A.; Zheng, S.J.; Van Loon, J.J.A.; Pieterse, C.M.J.; Dicke, M. Helping plants to deal with insects: The role of beneficial soil-borne microbes. *Trends Plant Sci.* **2010**, *15*, 507–514. [CrossRef]
51. Jiméncz, D.R.; Yokomi, R.K.; Mayer, R.T.; Shapiro, J.P. Cytology and physiology of silverleaf whitefly-induced squash silverleaf. *Physiol. Mol. Plant Pathol.* **1995**, *46*, 227–242. [CrossRef]
52. Mayer, R.T.; Inbar, M.; Doostdar, H. Are Pathogenesis-related proteins involved in plant resistance to insects? *Am. Soc. Plant Biol. Oster Plant Interact. Organ.* **1997**, *114*, 1120.
53. Inbar, M.; Doostdar, H.; Mayer, R.T. Effects of sessile whitefly nymphs (Homoptera: Aleyrodidae) on leaf-chewing larvae (Lepidoptera: Noctuidae). *Physiol. Chem. Ecol.* **1999**, *28*, 353–357. [CrossRef]
54. Van de Ven, W.T.G.; LeVesque, C.S.; Perring, T.M.; Walling, L.L. Local and systemic changes in squash gene expression in response to silverleaf whitefly feeding. *Plant Cell* **2000**, *12*, 1409–1423. [CrossRef] [PubMed]
55. Antony, B.; Palaniswami, M.S. *Bemisia tabaci* feeding induces pathogenesis-related proteins in cassava (*Manihotesculenta* Crantz). *Ind. J. Biochem. Biophys.* **2006**, *43*, 182–185.
56. Gorovits, R.; Akad, F.; Beery, H.; Vidavsky, F.; Mahadav, A.; Czosnek, H. Expression of stress-response proteins upon whitefly-mediated inoculation of Tomato yellow leaf curl virus in susceptible and resistant tomato plants. *Mol. Plant Microbe Interact.* **2007**, *20*, 1376–1383. [CrossRef] [PubMed]
57. Taggar, G.K.; Gill, R.S.; Gupta, A.K.; Sandhu, J.S. Fluctuations in peroxidase and catalase activities of resistant and susceptible black gram (*Vignamungo* (L.) Hepper) genotypes elicited by *Bemisia tabaci* (Gennadius) feeding. *Plant Sing. Behav.* **2012**, *7*, 1321–1329. [CrossRef] [PubMed]
58. Zhao, H.; Sun, X.; Xue, M.; Zhang, X.; Li, Q. Antioxidant enzyme responses induced by whiteflies in tobacco plants in defense against aphids: Catalase may play a dominant role. *PLoS ONE* **2016**, *27*, e0165454. [CrossRef]
59. Soliman, A.M.; Idriss, M.H.; El-Meniawi, F.A.; Rawash, I.A. Induction of pathogenesis-related (PR) proteins as a plant defense mechanism for controlling the cotton whitefly *Bemisia tabaci*. *Alex. J. Agric. Sci.* **2019**, *64*, 107–122. [CrossRef]
60. Graham, L.S.; Sticklen, M.B. Plant chitinases. *Can. J. Bot.* **1994**, *72*, 1057–1083. [CrossRef]
61. Singh, R.; Upadhyay, S.K.; Singh, M.; Sharma, I.; Sharma, P.; Kamboj, P.; Saini, A.; Voraha, R.; Sharma, A.K.; Upadhyay, T.K.; et al. Chitin, chitinases and chitin derivatives in biopharmaceutical, agricultural and environmental perspective. *Biointerface Res. Appl. Chem.* **2021**, *11*, 9985–10005.

62. Lawrence, S.D.; Novak, N.G. Expression of poplar chitinase in tomato leads to inhibition of development in Colorado potato beetle. *Biotechnol. Lett.* **2006**, *28*, 593–599. [CrossRef]
63. Castro, D.; Torres, M.; Sampedro, I.; Martínez-Checa, F.; Torres, B.; Béjar, V. Biological Control of verticillium wilt on olive trees by the salt-tolerant strain *Bacillus velezensis* XT1. *Microorganisms* **2020**, *8*, 1080. [CrossRef]
64. Sherzad, Z.; Canming, T. A new strain of *Bacillus velezensis* as a bioagent against *Verticillium dahliae* in cotton: Isolation and molecular identification. *Egypt. J. Biol. Pest Control* **2020**, *30*, 118. [CrossRef]
65. Ramiro, D.A.; Guerreiro-Filho, O.; Mazzafera, P. Phenol contents, oxidase activities, and the resistance of coffee to the leaf miner *Leucoptera coffeella*. *J. Chem. Ecol.* **2006**, *32*, 1977–1988. [CrossRef] [PubMed]
66. Marusek, C.M.; Trobaugh, N.M.; Flurkey, W.H.; Inlow, J.K. Comparative analysis of polyphenol oxidase from plant and fungal species. *J. Inorg. Biochem.* **2006**, *100*, 108–123. [CrossRef] [PubMed]
67. Constabel, C.P.; Barbehenn, R. Defensive roles of polyphenol oxidase in plants. In *Induced Plant Resistance to Herbivory*; Springer: Dordrecht, The Netherlands, 2008; pp. 253–270.
68. Schuman, M.C.; Baldwin, I.T. The layers of plant responses to insect herbivores. *Annu. Rev. Entomol.* **2016**, *61*, 373–394. [CrossRef]
69. Summers, C.B.; Felton, G.W. Prooxidant effects of phenolic acids on the generalist herbivore *Helicoverpa zea* (Lepidoptera: Noctuidae): Potential mode of action for phenolic compounds in plant anti-herbivore chemistry. *Insect Biochem. Mol. Biol.* **1994**, *24*, 943–953. [CrossRef]
70. Mishra, B.B.; Gautam, S. Polyphenol oxidases: Biochemical and molecular characterization, distribution, role and its control. *Enzym. Eng.* **2016**, *5*, 2329–6674.
71. Erb, M.; Reymond, P. Molecular interactions between plants and insect herbivores. *Annu. Rev. Plant Biol.* **2019**, *70*, 527–557. [CrossRef]
72. Castañera, P.; Steffens, J.C.; Tingey, W.M. Biological performance of Colorado potato beetle larvae on potato genotypes with differing levels of polyphenol oxidase. *J. Chem. Ecol.* **1996**, *22*, 91–101.
73. Bhonwong, A.; Stout, M.J.; Attajarusit, J.; Tantasawat, P. Defensive role of tomato polyphenol oxidases against cotton bollworm (*Helicoverpa armigera*) and beet armyworm (*Spodoptera exigua*). *J. Chem. Ecol.* **2009**, *35*, 28–38. [CrossRef]
74. Mahanil, S.; Attajarusit, J.; Stout, M.J.; Thipyapong, P. Overexpression of tomato polyphenol oxidase increases resistance to common cutworm. *Plant Sci.* **2008**, *174*, 456–466. [CrossRef]
75. Zhu-Salzman, K.; Luthe, D.S.; Felton, G.W. Arthropod-inducible proteins: Broad spectrum defenses against multiple herbivores. *Plant Physiol.* **2008**, *146*, 852–858. [CrossRef] [PubMed]
76. Duffey, S.S.; Stout, M.J. Anti-ntritive and toxic components of plant defense against insects. *Arch. Insect Biochem. Physiol.* **1996**, *32*, 3–37. [CrossRef]
77. Prabhukarthikeyan, S.R.; Keerthana, U.; Archana, S.; Raguchander, T. Induced resistance in tomato plants to *Helicoverpa armigera* by mixed formulation of *Bacillus subtilis* and *Beauveria bassiana*. *Res. J. Biotechnol.* **2017**, *12*, 53–59.
78. Senthilraja, G.; Anand, T.; Kennedy, J.S.; Raguchander, T.; Samiyappan, R. Plant growth promoting rhizobacteria (PGPR) and entomopathogenic fungus bioformulation enhance the expression of defense enzymes and pathogenesis-related proteins in groundnut plants against leaf miner insect and collar rot pathogen. *Physiol. Mol. Plant Pathol.* **2013**, *82*, 10–19. [CrossRef]
79. Barbehenn, R.; Dukatz, C.; Holt, C.; Reese, A.; Martiskainen, O.; Salminen, J.P.; Yip, L.; Tran, L.; Constabel, C.P. Feeding on poplar leaves by caterpillars potentiates foliar peroxidase action in their guts and increases plant resistance. *Oecologia* **2010**, *164*, 993–1004. [CrossRef]
80. Dowd, P.F.; Lagrimini, L.M. Examination of the biological effects of high anionic peroxidase production in tobacco plants grown under field conditions. I. Insect pest damage. *Transgenic Res.* **2006**, *15*, 197–204.
81. Qiu, Y.; Yan, H.; Sun, S.; Wang, Y.; Zhao, X.; Wang, H. Use of *Bacillus velezensis* SDTB022 against tobacco black shank (TBS) and the biochemical mechanism involved. *Biol. Control* **2022**, *165*, 104785. [CrossRef]
82. Kloepper, J.W.; Ngumbi, E.N.; Nangle, K.W.; Fadamiro, H.Y. Inoculants Including Bacillus Bacteria for Inducing Production of Volatile Organic Compounds in Plants. U.S. Patent EA201390851A1, 30 January 2014. pp. 2025–2037.
83. Disi, J.O.; Zebelo, S.; Kloepper, J.W.; Fadamiro, H. Seed inoculation with beneficial rhizobacteria affects European corn borer (Lepidoptera: Pyralidae) oviposition on maize plants. *Entomol. Sci.* **2018**, *21*, 48–58. [CrossRef]
84. Falqueto, S.A.; Pitaluga, B.F.; de Sousa, J.R.; Targanski, S.K.; Campos, M.G.; Mendes, T.A.O.; Silva, F.G.; Silva, D.H.S.; Soares, M.A. *Bacillus* spp. metabolites are effective in eradicating *Aedes aegypti* (Diptera: Culicidae) larvae with low toxicity to non-target species. *J. Invertebr. Pathol.* **2021**, *179*, 107525. [CrossRef]
85. Rajendran, L.; Ramanathan, A.; Durairaj, C.; Samiyappan, R. Endophytic *Bacillus subtilis* enriched with chitin offer induced systemic resistance in cotton against aphid infestation. *Arch. Phytopathol Plant Prot.* **2011**, *44*, 1375–1389. [CrossRef]
86. Hanafi, A.; Traore, M.; Schnitzler, W.H.; Woitke, M. Induced resistance of tomato to whiteflies and *Phytium* with the PGPR *Bacillus subtilis* in a soilless crop grown under greenhouse conditions. *Acta Hortic.* **2007**, *747*, 315–322. [CrossRef]
87. Valenzuela-Soto, J.H.; Estrada-Hernández, M.G.; Ibarra-Laclette, E.; Délano-Frier, J.P. Inoculation of tomato plants (*Solanum lycopersicum*) with growth-promoting *Bacillus subtilis* retards whitefly *Bemisia tabaci* development. *Planta* **2010**, *231*, 397–410. [CrossRef] [PubMed]

Article

Farmers' Knowledge on Whitefly Populousness among Tomato Insect Pests and Their Management Options in Tomato in Tanzania

Secilia E. Mrosso [1,2,3,*], Patrick Alois Ndakidemi [1,2] and Ernest R. Mbega [1,2]

[1] Nelson Mandela African Institution of Science and Technology, School of Life Science and Bioengineering, Arusha P.O. Box 447, Tanzania
[2] Centre for Research, Agricultural Advancement, Teaching Excellence and Sustainability in Food and Nutritional Security (CREATES), Nelson Mandela African Institution of Science and Technology, Arusha P.O. Box 447, Tanzania
[3] Tanzania Plant Health and Pesticides Authority, Arusha P.O. Box 3024, Tanzania
* Correspondence: smrosso@yahoo.com or secilia.mrosso@tphpa.go.tz

Abstract: Whitefly is a populous insect pest among tomato insect pests, causing significant crop loss through direct and indirect attacks. The current study aimed to assess the knowledge of tomato farmers on the populousness of whiteflies compared to other tomato insect pests and explore the management options available in their farming context in three tomato-growing regions, Arusha, Morogoro, and Iringa, in Tanzania. The study used a questionnaire to collect the data with backup information obtained through key informants' interviews and focus group discussions. The study findings indicated whitefly to be populous among tomato insect pests. However, tomato farmers showed varying knowledge of whitefly aspects, including differing control options for the pest. Such findings indicated a knowledge gap between farmers' understandings of the pest and their practices in fighting it compared to the standard and required practices in controlling the pest.

Keywords: whitefly populousness; knowledge; farmers' perception; pest control and pesticides application skills

Citation: Mrosso, S.E.; Ndakidemi, P.A.; Mbega, E.R. Farmers' Knowledge on Whitefly Populousness among Tomato Insect Pests and Their Management Options in Tomato in Tanzania. *Horticulturae* 2023, 9, 253. https://doi.org/10.3390/horticulturae9020253

Academic Editors: Eligio Malusà and Małgorzata Tartanus

Received: 26 December 2022
Revised: 27 January 2023
Accepted: 6 February 2023
Published: 13 February 2023

Copyright: © 2023 by the authors. Licensee MDPI, Basel, Switzerland. This article is an open access article distributed under the terms and conditions of the Creative Commons Attribution (CC BY) license (https://creativecommons.org/licenses/by/4.0/).

1. Introduction

Whitefly (*Bemisia tabaci*-Gennadius) is a devastating insect pest of tomatoes in all production systems worldwide [1]. The insect is very polyphagous, affecting both cultivated and weed plant species [2]. Whitefly causes substantial crop losses through direct and indirect effects on the host plants [3]. Adults and nymph directly suck the plant phloem sap, which is rich in nitrogen in the form of free amino acids, soluble proteins, and soluble carbohydrates, causing significant nutrient competition among the host plant and the insect pest [4]. Indirectly, whiteflies vector more than 350 pathogenic plant viruses that threaten the production of tomatoes and other crops in the tropics and subtropics [5] and impair the trade of agricultural commodities [6]. The most prevalent virus that affects tomato production are tomato yellow leaf curl virus (TYLCV) from the genus Begomovirus which cause tomato yellow leaf curl disease (TYLCD) in the tropics and subtropics, Tanzania included [7]. The virus is the most devastating among the Begomovirus, threatening tomato production globally [8]. This virus is very common and widespread in Tanzania, especially in a hot summer, where it can lead to 100% of tomato losses [9]. Its prevalence is high in Dodoma, Morogoro, Dar Es Salaam, Iringa Kilimanjato, and Arusha, as reported by the same study. Tomato mosaic virus is also a threat to tomato production in many parts of the world [10].

Whitefly is very destructive among tomato insect pests, causing tomato losses of up to 100%, which amounts to more than one hundred million dollars each year [11]. For example, in Nepal, whiteflies were reported to be populous and the leading insect pest in tomato production [12]. The loss is counted from the costs involved in purchasing the pesticides and other control measures, time spent applying the control measures, and the crop loss due to the insect pest attack in quantity and quality.

Most farmers use synthetic pesticides to control whiteflies as the preferred pest control method due to their fast action and mass insect-killing manner [13]. They also mix different pesticides to increase their synergy [13]. Additionally, tomato farmers use biological, cultural, and mechanical methods to control whiteflies [14]. Despite all these efforts, the whitefly insect pest continues to dominate tomato production in tropical and subtropical areas [15] where Tanzania is inclusive. This failure is anticipated from whitefly colonization strategies, such as the ability of whiteflies to hide under the leaf surface of the host plant [2], wide genetic diversity [16], wide host range [17,18], small body size and short life cycle [19], the ability to develop insecticides resistance, and high invasiveness [20], which give whiteflies advantages to survive the applied control methods.

To have effective whitefly control, apart from having the control measures in place, the awareness of farmers on whitefly, including their understanding of it as a pest and the wise selection and use of a control measure, is important. Therefore, farmers' awareness of whitefly, major tomato insect pests, control options available, and the use of pesticides to mitigate them with emphasis on the control methods, type of pesticide used, pesticide application frequency, and their perception of the best practices need to be assessed. As such, the current study was carried out to determine the knowledge of tomato farmers on whitefly populousness among the major tomato insect pests and explore the options for their management in Tanzania.

2. Materials and Methods
2.1. Study Sites and Data Collection

The study on the assessing of the tomato farmers' knowledge and practices against whiteflies (*Bemisia tabaci*-Gennadius) in Tanzania was conducted from June to September 2022. The study used a semi-structured questionnaire with both open- and closed-ended questions. Purposive sampling was used to select regions that are core tomato producers in the country where Arusha (3.3869° S, 36.6830° E), Morogoro (6.8278° S, 37.6591° E), and Iringa (7.7681° S, 35.6861° E) were chosen as indicated in Figure 1. Then, purposive sampling was carried out to select one district from each region, where Meru (3.4470° S, 36.6741° E), Mvomero (6.2555° S, 37.5535° E), and Kilolo (7.8835° S, 36.0893° E) districts were selected, respectively. Finally, respondents who are tomato farmers were selected randomly from tomato farmers of the three districts, where 50 respondents were chosen per district, making a total of 150 tomato farmers. These respondents were a representative sample of 62,663 households reported to engage in tomato production in Tanzania mainland [21].

The questionnaire was pre-tested on ten tomato farmers from the Meru district in the Arusha region, where the farming context is similar to the study area. Before starting the interview, we sought consent from the respondents by providing them with a form that introduced and explained the aim of the research and they were asked for approval to continue with the interview. Then, the questionnaire was administered to the respondents to enquire about their understanding and awareness of whiteflies and if it is a common pest in their area, its damaging stage and peak population during tomato growing season. Additionally, respondents were enquired to provide information related to the whitefly control strategies, such as the whitefly control methods they use, the most effective methods, pesticide usage, pesticide products used, and pesticide application frequency in their area (Table 1). Finally, an informants' interview and focus group discussion was conducted through an organized community meeting to supplement the data provided by the respondents. During the process, guiding questions were asked to provoke discussion regarding tomato production challenges and the management option highlighting the most effective methods.

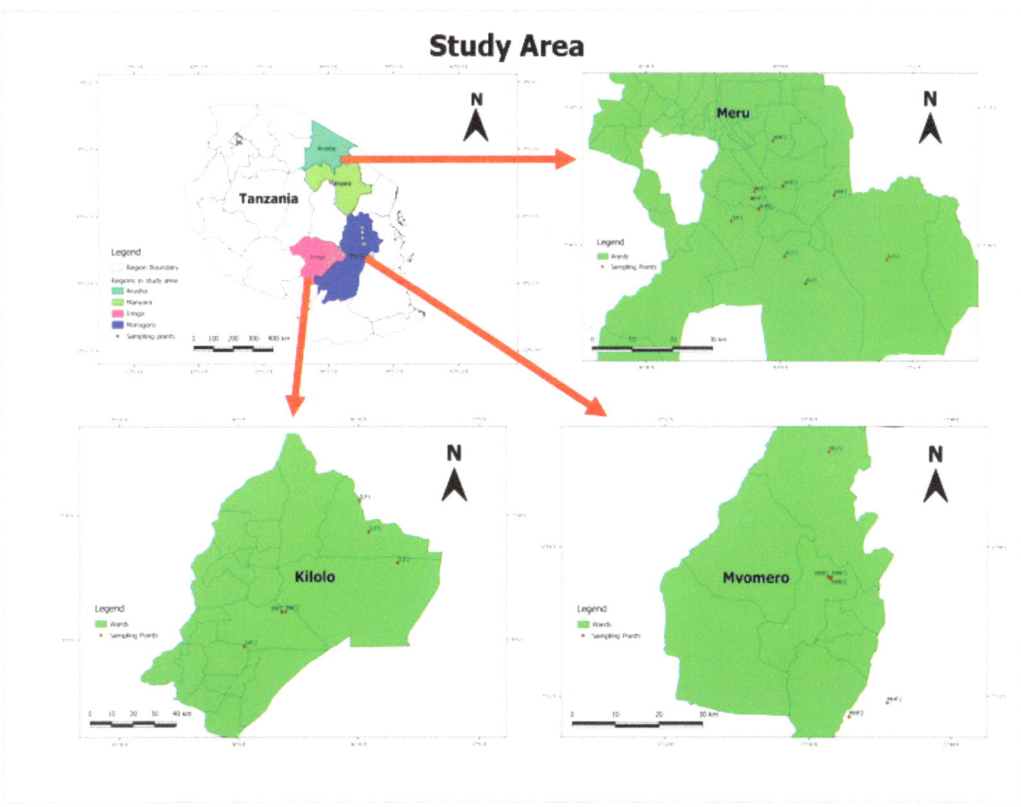

Figure 1. Map of Tanzania showing the study area.

Table 1. Overview of the questions included in the questionnaire used to assess the farmer's knowledge on whiteflies populousness among tomato insect pests and their management options in tomato in Tanzania.

Data Group	Description
Respondents' demographic data and farm characteristics	Gender, age, marital status, education, farm size, yield
Farmers' knowledge and perception of tomato pests	Common tomato production problems, critical tomato production problems, a common insect pest of tomato, if tomato producers are aware of whitefly, whether whitefly is a common insect pest in the respondent's area, destruction stage of whitefly, damage symptoms of whiteflies and whitefly peak time, perceptions of the impact of whitefly on tomato yields and whether whitefly is a populous insect pest in tomato production in the respondents' area.
Whitefly management practices perception	Whitefly control methods, control method that works better, pesticide use; pesticide products; pesticide spraying frequency in the field

2.2. Data Analysis IBM SPSS Statistics for Macintosh, Version 25.0

The statistical package for social science software - IBM SPSS Statistics (Version 16) summarized the survey data into descriptive statistics, such as percentages and means. Percentages were calculated for each group of similar responses from multiple answered questions. A Chi-square was used to assess the differences in knowledge and perception of the respondents on tomato pests, including whiteflies and their management practices.

3. Results

The outcome of this study shed light on the farmers' understanding of the best methods of whitefly control, whitefly peak time within the production season, the damaging stage, and symptoms that will assist in developing whitefly management approaches for increased tomato production in Tanzania, as discussed hereunder.

3.1. Distribution of Respondents across Demographic Variables

The study explores the demographic characteristics of the study respondents as they contribute to their perception of various life aspects. Details of each variable are presented in Table 2.

Table 2. Socio-economic characteristics of the study respondents and their tomato yield.

Variable		Frequency	Percentage
Gender	Male	116	77.3
	Female	34	22.7
	Total	150	100
Age	15–24	4	2.7
	25–34	24	16.0
	35–44	69	46.0
	45–55	37	24.7
	55–64	14	9.3
	65+	2	1.3
	Total	150	100
Marital status	Married	133	88.7
	Single	17	11.3
	Total	150	100
Education level	Primary	121	82.0
	Secondary	21	14.0
	Tertiary	8	4.0
	Total	150	100
Farm size in Ha	0.2–0.4	99	66.7
	0.6–0.8	42	28.7
	>0.8	9	4.6
	Total	150	100
Farming experience in years	2	31	20.7
	4	57	38.0
	5+	62	41.3
	Total	150	100
Tomato harvest/Ha	6–9	9	6.1
	9.5–11	23	15.3
	11.5–15	62	41.3
	15.5–19	56	37.3
	Total	150	100.0

3.2. Farmers' Knowledge and Perception of Tomato Production Problems

Respondents of the study had diverse knowledge and perceptions of aspects relating to tomato pests. When it comes to the variety of tomatoes cultivated, most of them (54.7%) grow hybrid tomato varieties. On the other hand, 25.3% grow hybrid and open-pollinated tomato varieties, while 20% grow open-pollinated tomato varieties.

Furthermore, the respondent reported different problems that they faced during tomato production, as indicated in Table 3. Of the 150 respondents, 71.3% reported insect pests, while 16% reported insect pests and diseases. Bad weather was reported by 6% of the respondents, while 5.3% reported diseases as a barrier to tomato production. Only 1.3% reported poor soil fertility as a tomato production problem. Respondents were asked if they were aware of whiteflies as an insect pest in tomato production, where about 74.7% and 25.3% reported yes and no, respectively. Since insect pests were mentioned as a major tomato production problem, the study asked for common tomato insect pests. Whiteflies ranked first, occupying 65.3% of the respondents. Tomato leaf miner and American ball worm were reported by 21.3% of the respondents, while 12.1% and 1.3% reported Tomato leaf minor and American ball worm, respectively.

Table 3. Farmers' knowledge and perception on various aspects of tomato pests in the study area.

	Variable	Frequency	Percentage
	OPV	30	20.0
Tomato varieties cultivated	Hybrid	82	54.7
	OPV and Hybrid	38	25.3
	Total	150	100
	Insect pest	107	71.3
	Diseases	8	5.3
Common tomato production problems	Bad weather	9	6.0
	Poor soil fertility	2	1.3
	Insect pests and disease	24	16.0
	Total	150	100
	Yes	112	74.7
If the respondent is aware of whiteflies	No	38	25.3
	Total	150	100
	Tomato leaf miner	18	12.1
Common insect pests of tomato in the area of respondent	American Ball worm	2	1.3
	Whiteflies	98	65.3
	Tomato leaf miner and American ball worm	32	21.3
	Total	150	100
	Adult	100	66.7
Destruction stage of whiteflies	Nymph	27	18.0
	Both adult and nymph	23	15.3
	Total	150	100
	Leaf yellowing and curling	61	40.7
	Plant stunting	26	17.3
Whiteflies damage symptoms	Plant wilting	31	20.7
	Do not know	32	21.3
	Total	150	100
	In the nursery	1	0.7
Peak whiteflies population in tomato growing season	The first month after transplanting	36	24.0
	Flowering stage	61	40.7
	All the production season	52	34.7
	Total	150	100

Respondents were asked if they were aware of the destruction stage of whiteflies, and 66.7% and 18% pointed out adult and nymph, respectively. Destruction by both nymph and adult whiteflies were mentioned by 15.3% of the respondents. Whitefly damaging symptoms were also a question of interest, of which 40.7% of the respondents pointed out leaf yellowing and curling while 20.7% said plant wilting. Plant stunting was mentioned by 17.3%, and 21.3% had no idea about the whitefly damage symptoms. Additionally, the peak whiteflies population in tomato growing season was inquired, and 40.7% and 34.7% responded during the flowering stage and all of the production season, respectively. Another 24% reported that it is in the first month after transplanting, while 7% reported it in the nursery.

3.3. Whitefly Management Practices

The study also enquired about the knowledge and perception of the respondents on the whitefly management practices in their areas as indicated in Table 4. Their responses were distributed such that 78.7% perceived the chemical method and 10% the field and surroundings sanitation. About 9.3% perceived using integrated pest management (IPM), and 2% used cultural practices. The response on the respondents' perception of whitefly control methods that work better in their farming context was as follows: 82.7% perceived the chemical whitefly control method to be better, while 9.3% and 8% perceived cultural and IPM methods to be better, respectively. The pesticide application knowledge of respondents was enquired about as it affects the performance of a particular pesticide. The responses indicated only 15.3% know, with the majority, 84.7%, applying pesticides with no pesticide application knowledge, which may have contributed to the populousness of whiteflies in tomato farming in Tanzania.

Table 4. Farmers' perception on various aspects of whitefly management practices.

	Variable	Frequency	Percentage
Respondents Whitefly management practices	Chemical method	118	78.7
	Cultural method	3	2.0
	IPM	14	9.3
	Field and surroundings sanitation	15	10.0
	Total	**150**	**100.0**
Whitefly management option(s) that work(s) better	Chemical method	124	82.7
	Cultural method	14	9.3
	IPM	12	8.0
	Total	**150**	**100.0**
If respondents have pesticide application knowledge	Yes	23	13.3
	No	127	84.7
	Total	**150**	**100**
If respondents use synthetic pesticides to control whiteflies	Yes	124	82.7
	No	26	17.3
	Total	**150**	**100.0**
Type of synthetic pesticide a respondent use	Snow tiger-Chlorfenapyr10%	26	17.3
	Snow thunder-Thiamethoxam3% + Emamectin Benzoate 1%	41	27.3
	Profecron- Profecros750G/L	22	14.7
	Dudu will—Cypermethrin	20	13.3
	Snow thunder and snow tiger	18	12.0
	Profecron and snow tiger	20	13.3
	Snow tiger and Duduwill	3	2.0
	Total	**150**	**100.0**
Source of extension services in tomato production	Government	64	42.7
	Private	56	37.3
	Both government and private	30	20.0
	Total	**150**	**100.0**
Guidance on the pesticide rate of application	As per label instruction	74	49.3
	Experience	22	14.7
	As per the extension officer's advice	31	20.7
	None	23	15.3
	Total	**150**	**100.0**
Frequency of pesticide application in the tomato growing season	Once	33	22.0
	Twice	8	5.3
	Three times	61	40.7
	More than three times	26	17.3
	None	22	14.7
	Total	**150**	**100.0**
Amount of crop loss due to whiteflies infestation if not controlled	Total crop loss	65	43.3
	20%	7	4.7
	More than 20%	48	32.0
	Do not know	30	20.0
	Total	**150**	**100.0**

On the type of synthetic pesticide used by a respondent to control whitefly, 27.3% used Snow thunder (Thiamethoam3% + Emamectin Benzoate 1%) and 17.3% used Snow tiger (Chlorfenapyr10%). About 14.7% used Profecron (Profecros750G/L), and 13.3% and another 13.3% used Dudu will (Cypermethrin and Profecron) and snow tiger, respectively. About 12% used Snow thunder and Snow tiger, while a small proportion of the respondents (2%) used Snow tiger and Dudu will. Respondents were also asked from which source they received extension services during tomato production, where 42.7% received assistance from the government, 37.3% from private sector, while 20% received it from both of the two sources. On whom guided them on the pesticide rate of application, their responses were such that 49.3% were by the pesticide label instruction, and 20.7% received extension services that guided them on the matter. About 14.7% apply pesticides based on their experience, while 15.3% did not use any guide in determining the pesticide application rates. Pesticide application frequency during the tomato production season was such that 22% of respondents applied pesticide once, while 5.3% applied it twice, 40.7% applied it three times, and 17.3% applied more than three times. About 14.7 did not use pesticides to control whiteflies during tomato production. The respondents' views on the amount of crop loss due to whitefly attack were very diverse. Most of respondents (43/3%) reported a total loss, while 4.7% reported a loss of 20%. A total of 32% of respondents reported a loss of more than 20%, while 20% of respondents could not estimate the amount of crop loss in this regard, tomatoes, attributed to whiteflies.

3.4. Association among the Study Variables by Crosstabulation

The study variables were compared to determine whether they associate with each other. The variable relationship sheds light on how to deal with them as they influence each other or not. A crosstabulation between the level of education a respondent attained and the amount of tomato that the same individual harvested in tons per hectare was significant ($p = 0.000$). Therefore, these two variables are associated with each other, as in Table 5.

Table 5. Respondents' education level. Tomato yield (tons/Ha) crosstabulation.

			Tomato Harvest				Total
			6–9	9.5–11	11.5–15	15.5–19	
Respondent education level	Primary	Count	14	34	58	15	121
		Expected Count	11.3	29.0	52.4	28.2	121.0
	Secondary	Count	0	1	6	14	21
		Expected Count	2.0	5.0	9.1	4.9	21.0
	Tertiary	Count	0	1	1	6	8
		Expected Count	0.7	1.9	3.5	1.9	8.0
Total		Count	14	36	65	35	150
		Expected Count	14.0	36.0	65.0	35.0	150.0

$X^2 = 43.54$, df = 6, $p = 0.000$

There was also an association between the number of years a respondent has been in tomato production and the number of tomatoes harvested with $p = 0.000$ (Table 6).

Further, an association was shown between the farmers' age in years and the tomato farming experience one has accumulated with $p = 0.000$ (Table 7).

Table 6. Tomato farming experience in years. Tomato harvest crosstabulation.

			Tomato Harvest				Total
			6–9	9.5–11	11.5–15	15.5–19	
Years you have been in tomato production	2	Count	3	9	12	7	31
		Expected Count	1.7	5.0	12.7	11.7	31.0
	4	Count	3	12	29	13	57
		Expected Count	3.1	9.2	23.3	21.4	57.0
	5+	Count	2	3	20	36	61
		Expected Count	3.3	9.8	25.0	22.9	61.0
Total		Count	8	24	61	56	149
		Expected Count	8.0	24.0	61.0	56.0	149.0

$X^2 = 25.4$, df = 6, $p = 0.000$

Table 7. Farmers' age in years. Tomato farming experience in years crosstabulation.

			Years You Have Been in Tomato Production			Total
			2	4	5+	
Age of respondent	15–24	Count	4	0	0	4
		Expected Count	0.8	1.5	1.6	4.0
	25–34	Count	18	6	0	24
		Expected Count	5.0	9.2	9.8	24.0
	35–44	Count	9	47	12	68
		Expected Count	14.1	26.0	27.8	68.0
	45–54	Count	0	4	33	37
		Expected Count	7.7	14.2	15.1	37.0
	55–64	Count	0	0	14	14
		Expected Count	2.9	5.4	5.7	14.0
	65 and above	Count	0	0	2	2
		Expected Count	0.4	0.8	0.8	2.0
Total		Count	31	57	61	149
		Expected Count	31.0	57.0	61.0	149.0

$X^2 = 1.47$, df = 10, $p = 0.000$

4. Discussion

The current study aimed to gauge farmers' knowledge and practices in managing whiteflies in tomato production in Tanzania. Demographic characteristics of the studied population were important in the study. Study results in Table 2, showed that most of land ownership and allocation in various production activities seem to be determined by gender, where men were more favored. As a result, women are discriminated against in land ownership and utilization, despite contributing to 52% of the agricultural labor force in Tanzania [22]. This discrimination affects their decision to engage in farming and their perception of agriculture in general. However, reports narrate that less than 15% of landholders worldwide are women [23]. Suppose there could be equality in land ownership among men and women; women as a key provider of the farming labor force could be encouraged to devote more energy to the sector. In one way or another, this could affect their perception of farming aspects, including fighting crop insect pests. Other studies conducted in the Mvomero district in the Morogoro region and in the Musoma municipality also reported men to dominate tomato production [24,25]. Men are reported to have more access to capital, therefore have more power to fund tomato production activities, as it is labor intensive [26].

Age is another factor influencing one's perception of farming as it relates to a person's experience with something. In the current study, most respondents are aged 15–64, the active labor force age in Tanzania [23]. Such a labor force is expected to actively participate in the study. Studies reported a positive correlation between age and the efficiency of economic inputs [23]. The same age group was reported to the active group engaging in crop production [26]. On top of that, most study respondents with active age had four and above years of experience in tomato farming ($X^2 = 1.47$, df = 10 and $p = 0.000$) that can allow them to take an active part in economic activities, as indicated in Table 7. The education of respondents varied significantly. However, all respondents could read and write as they had a reasonable formal education ranging from primary to tertiary. Such an ability can influence the respondents' perception of tomato pests and the decision to choose and apply a control measures, as education is a determinant factor in the adoption of innovations [27]. It also influences their resource allocation in tomato production, such as land and yield, as revealed in this study.

All farmers under the study were small-scale producers, as the majority had farms at most 2 ha [28]. The study findings align with other studies that reported tomato production on farm sizes of 0.56 Ha and 0.4 Ha in Morogoro and Kenya, respectively [26,29]. However, these farmers have accumulated enough farming experience, as most have been in farming for four and more than five years. Such accumulated farming experience helped farmers to have more tomato yield/Ha, as further indicated by the crosstabulation results between farming experience and the tomato yield obtained in tons/Ha, shown in Table 6. The results align with another study that reported that farmers' characteristics influence their farming behavior [30].

4.1. Farmers' Knowledge and Perception of Different Aspects of Tomato Pests

Results from Table 3 indicated farmers differ in their choice of tomato variety selection whether hybrid or open-pollinated varieties (OPV). However, the majority selected hybrids, as they are bred for specific qualities through plant breeding, such as pest resistance, and therefore are preferred over OPVs due to their ability to resist various production problems. The findings are in line with another study that reported that the hybrids of horticultural crops could tolerate environmental stresses [31]. Additionally, in Table 3, respondents reported problems they encounter during tomato production, with insect pest ranking first. These study findings align with other studies that reported insect pests as the major threat to crop production [32,33].

Further, respondents outlined a list of insect pests facing tomato production in their farming context. Whiteflies occupied the largest share among tomato insect pests, ranking first in the list, indicating whiteflies to be populous and a big problem in tomato production in the country. This understanding also indicates that respondents were aware of whitefly. The same insect pest was reported to threaten tomato production worldwide [2] The populousness of whitefly as an insect pest was also reported in Nepal [12].

Other insect pests of importance to tomatoes were leaf minor and American ball worm, as reported by [34]. Interview results of the key informants also outlined insects as a key problem in tomato production, emphasizing whitefly as a threat to tomato production.

The fact that all the respondents possessed a formal education and most of them are in the active labor force group may have contributed to their ability to identify the tomato pests precisely. However, regarding the whitefly destruction stage, respondents need knowledge on the same as most need to learn exactly what stage in the whitefly lifecycle is destructive. Actually, both the nymph and adults are responsible for the host plant destruction [35].

Awareness of a pest's destructive stage helps target it at the right time and shed light on the right control means. Therefore, it is important to facilitate farmers' understanding of the matter. In connection to understanding the whitefly destruction stage, understanding the damaging symptoms caused by whitefly is also important as it sharpens the focus on applying the pest control method. Most respondents did not understand this aspect as they could not point out all the damaging whitefly symptoms. A plant attacked by whiteflies will develop symptoms such as leaf yellowing and curling, plant wilting, and plant stunting, and impaired fruit ripening. Such signs were also reported by other studies [36]. Respondents have differing perceptions of the peak whiteflies population in the tomato growing season. Their responses varied greatly indicating their need for more awareness. The whitefly population starts progressively from the nursery and continues to grow from transplanting, the first month of the crops in the field. As the crops progress in the field, the whitefly population grows larger if uncontrolled. At the fruit setting stage, this insect's population is as large as reaching the economic threshold, where the cost of applying control measures cannot justify the crop recovered. Tomato producers must understand this to use the control measure before this time. It is even more alarming as studies reported whitefly as a disastrous insect pest worldwide [37,38], where only one adult/leaf and four nymph/leaf are enough to cause mediated economic injury that calls for control measure application [39].

4.2. Whitefly Management Practices

Whitefly management practices are measures against whiteflies. They vary among individuals within the farming context of a particular area. The same scenario happened when the study respondents used varying whitefly control measures due to their different perceptions of which method works better, as in Table 4. From this study, whitefly control methods applied by the respondents varied from chemical pesticides to field and surroundings sanitation, cultural practices, and integrated pest management (IPM). Most respondents used the chemical method, and backup information from the focus group discussion mentioning chemical pesticides as the main method of pest control in their areas. This information agrees with other studies that reported chemical pesticides as the main and first bullet in dealing with insect pests in crop production [40,41]. The reasons for the first choice of pesticides may be due to pesticides' ability to kill many insects within a short period making the method convenient and highly effective [42]. The respondent's ability to read and write, as determined by their education level, may also have guided them in the selection of this method of pest control. Additionally, the experience the farmers build through engagement in tomato production for more than four years for most of them is an added advantage selecting pest control methods. Experience is said to be the best teacher, as in Table 2. However, chemical insect pest control is said to be ineffective, as whiteflies can develop pesticide resistance, especially when a pesticide with a single mode of action is used repeatedly [40]. Other methods that showed promise in controlling whiteflies were cultural and IPM methods, as also found effective in controlling the same pest in other places [43]. Some respondents outlined cultural practices: crop rotation, intercropping, and proper fertilizer usage and irrigation regulation.

Tomato farmers used a variety of synthetic pesticides in controlling whiteflies either singly or even in combination by mixing several chemical molecules aiming at increasing chemical synergies. However, the case may differ as the farmers need to gain knowledge of pesticide application techniques [13], which may include a low knowledge of pesticide compatibility. Therefore, mixing pesticides may accelerate their harmful effects on the environment and the ecosystem in general while raising the production costs through the purchase of the pesticide, application time, and the loss of the crop produced due to pesticide contamination or destruction by the insect pest as a result of loss of pesticide effectiveness [44]. Pesticide mixing in an attempt to control insect pests in crop production, particularly whiteflies, was also reported by another study [45]. Still on the contrary, they were not effective on the target. The farmers' decision to mix insecticides with no prior information on the effectiveness of the resultant product and its effect on the environment and the non-target organisms may be attributed to the low level of education possessed by most of the respondents, as indicated in Table 2. Similar results were reported by other studies as well [14,46].

Furthermore, the source of farmers' guidance on proper pesticide usage differed, as shown in Table 4. However, a very small proportion of farmers (20.7) used pesticides based on extension advice. Lack of understanding of the proper pesticide usage, and the primary school education level possessed by most farmers, can compromise the quality of the farmers' practices, such as the pesticide application rate used. Pesticides are harmful substances, and their handling and application require a prior understanding of their side effects when mishandled. Additionally, a lack of guidance renders pesticides less effective and increases the chances of insect pests developing pesticide resistance, making them even more difficult to manage [14]. However, the government is the main source of extension services, where the services are less effective as they are not delivered in a timely manner due to the dispersed nature of rural farmers. Such a scenario necessitates the joined efforts of both the public and private sectors to raise the farmers' knowledge on safe pesticide usage. These findings are in line with other studies reporting related results [47].

Pesticide application frequency within tomato growing indicated a dangerous pesticide application frequency trend, as in Table 4. Tomato growing season takes only three months, and pesticide application frequency of more than three times per season can lead to pesticide overuse which contaminates the environment and the crop produced while increasing the production costs on the farmers' side [48]. It also increases cha chances of the pest developing pesticide resistance to those particular pesticide molecules [44]. Additionally, a single pesticide application within the tomato growing season can be the underutilization of pesticide, which can also lead to ineffective control of the pests. Such discrepancies are attributed by most farmers (84.7%) to a lack of pesticide application knowledge, as reported in this study. For instance, the study findings in Table 5 reported a positive relationship between education level and the tomato yield obtained ($X^2 = 43.54$, df = 6 and $p = 0.000$). However, the relationship between the respondents' education level and the pesticides application rate was insignificant, as in Table 8, due to farmers relying on experience.

Table 8. Education level of the respondent. Rate of application of pesticides in tomato production crosstabulation.

Count							
			Rate of Application of Pesticides in Tomato Production				
			As per Label Instruction	Experience	As per Extension Officer's Advice	None	Total
Education level of the respondent		Primary	59	20	22	20	121
		Secondary	10	2	7	2	21
		Tertiary	5	0	2	1	8
Total			74	22	31	23	150

$X^2 = 4.82$, df = 6, $p = 0.567$

Pesticide application knowledge sheds light on application rate, method, frequency, stage of pest, and crop. It is, therefore, highly needed for the benefit of the farmers' health, economic benefit, and the betterment of the environment and the ecosystem in general.

Finally, the perception of farmers on the amount of crop losses due to whiteflies was different among them. Their responses varied from total loss to ≥20% failure, with others unable to tell the amount of loss attributed to whiteflies. However, studies reported a loss of up to 100% equating to a hundred million dollars per year [11]. The variation in these responses indicates the need for awareness creation to bring farmers' understanding to the same level in the fight against whiteflies.

5. Conclusions

The study on farmers' knowledge on whitefly populousness among tomato insect pests and their management options in tomato in Tanzania found out that tomato production in Tanzania is practiced by small-scale active-age farmers. The farmers' demographic characteristics are different among themselves, which in turn determined their knowledge on various aspects of whitefly control. The farmers possess differing knowledge on whitefly as a pest and on its control means.

Viewing the results from this angle, there is a need for imparting tomato farmers with knowledge on tomato production aspects, particularly insect pests. Their understanding of the whitefly damaging stage, the damaging symptoms, their peak population, and crop loss need to be updated. They also need a common understanding of the best whitefly control practices that are available in their farming context to practice them and have whiteflies controlled with minimal efforts for increased tomato production and productivity. Therefore, both the government and the private sector are called upon to partner to reach as many farmers as possible and impart them with the needed knowledge to economically benefit tomato production. Additionally, more research on whitefly control methods that seem to be the best is needed to keep the number of whiteflies low enough to mitigate their negative impacts on the crop and eventually on the economy of the tomato farmers and the country as a whole.

Author Contributions: Conceptualization, S.E.M., P.A.N. and E.R.M.; methodology, S.E.M.; formal analysis, S.E.M.; writing—original draft preparation, S.E.M.; writing—review and editing, P.A.N. and E.R.M. All authors have read and agreed to the published version of the manuscript.

Funding: This research was funded by the World Bank through the African Centre for Research, Agricultural Advancement, Teaching Excellence and Sustainability in Food and Nutrition Security (CREATES), in the School of Life Sciences and Bioengineering hosted at the Nelson Mandela African Institution of Science and Technology (NM-AIST), grant ID: P151847. The APC was funded by Secilia E. Mrosso, the corresponding author of this article.

Informed Consent Statement: Informed consent was obtained from all subjects involved in this study.

Data Availability Statement: The data used in writing this article are within the article/paper.

Acknowledgments: The authors thank the leaders in the department of Agriculture and Livestock of Arusha, Morogoro and the Iringa regions and the same leaders at the districts of Meru, Mvomero and Kilolo for allowing the authors to carry out the study in their areas.

Conflicts of Interest: The authors declare that they have no known competing financial interests or personal relationships that could have appeared to influence the work reported in this paper.

References

1. Cruz-Estrada, A.; Gamboa-Angulo, M.; Borges-Argáez, R.; Ruiz-Sánchez, E. Insecticidal effects of plant extracts on immature whitefly Bemisia tabaci Genn.(Hemiptera: Aleyroideae). *Electron. J. Biotechnol.* **2013**, *16*, 6.
2. Sri, N.R.; Jha, S. Whitefly biology and morphometry on tomato plants. *J. Entomol. Zool. Stud.* **2018**, *6*, 2079–2081.
3. Lee, M.-H.; Lee, H.-K.; Lee, H.-G.; Lee, S.-G.; Kim, J.-S.; Kim, S.-E.; Kim, Y.-S.; Suh, J.-K.; Youn, Y.-N. Effect of cyantraniliprole against of Bemisia tabaci and prevention of tomato yellow leaf curl virus (TYLCV). *Korean J. Pestic. Sci.* **2014**, *18*, 33–40. [CrossRef]
4. Žanić, K.; Dumičić, G.; Mandušić, M.; Vuletin Selak, G.; Bočina, I.; Urlić, B.; Ljubenkov, I.; Bučević Popović, V.; Goreta Ban, S. Bemisia tabaci MED Population Density as Affected by Rootstock-Modified Leaf Anatomy and Amino Acid Profiles in Hydroponically Grown Tomato. *Front. Plant Sci.* **2018**, *9*, 86. [CrossRef] [PubMed]
5. Ochilo, W.N.; Nyamasyo, G.N.; Kilalo, D.; Otieno, W.; Otipa, M.; Chege, F.; Karanja, T.; Lingeera, E.K. Ecological limits and management practices of major arthropod pests of tomato in Kenya. *J. Agric. Sci. Pract.* **2019**, *4*, 29–42. [CrossRef]
6. Gilbertson, R.L.; Batuman, O.; Webster, C.G.; Adkins, S. Role of the insect supervectors Bemisia tabaci and Frankliniella occidentalis in the emergence and global spread of plant viruses. *Annu. Rev. Virol.* **2015**, *2*, 67–93. [CrossRef] [PubMed]
7. Scholthof, K.B.G.; Adkins, S.; Czosnek, H.; Palukaitis, P.; Jacquot, E.; Hohn, T.; Hohn, B.; Saunders, K.; Candresse, T.; Ahlquist, P. Top 10 plant viruses in molecular plant pathology. *Mol. Plant Pathol.* **2011**, *12*, 938–954. [CrossRef] [PubMed]
8. Moriones, E.; Navas-Castillo, J. Tomato yellow leaf curl virus, an emerging virus complex causing epidemics worldwide. *Virus Res.* **2000**, *71*, 123–134. [CrossRef] [PubMed]
9. Kashina, B.D.; Mabagala, R.B.; Mpunami, A.A. Transmission properties of tomato yellow leaf curl virus from Tanzania. *J. Plant Prot. Res.* **2007**, *47*, 43–51.
10. Hussain, I.; Farooq, T.; Khan, S.; Ali, N.; Waris, M.; Jalal, A.; Nielsen, S.; Ali, S. Variability in indigenous Pakistani tomato lines and worldwide reference collection for Tomato Mosaic Virus (ToMV) and Tomato Yellow Leaf Curl Virus (TYLCV) infection. *Braz. J. Biol.* **2022**, *84*, e253605. [CrossRef] [PubMed]

11. Moodley, V.; Gubba, A.; Mafongoya, P.L.J.C.P. A survey of whitefly-transmitted viruses on tomato crops in South Africa. *Crop Prot.* **2019**, *123*, 21–29. [CrossRef]
12. Gauli, K.; Sah, L.; Shrestha, J.; Rajbhandari, B.; Ghimire, A. Major Insect Pests and Pesticide Use Practices among Tomato Growers in Kathmandu and Bhaktapur Districts. *J. Plant Prot. Soc.* **2020**, *6*, 202–211. [CrossRef]
13. Laizer, H.C.; Chacha, M.N.; Ndakidemi, P.A. Farmers' Knowledge, Perceptions and Practices in Managing Weeds and Insect Pests of Common Bean in Northern Tanzania. *Sustainability* **2019**, *11*, 4076. [CrossRef]
14. Mrosso, S.E.; Ndakidemi, P.A.; Mbega, E.R. Characterization of Secondary Metabolites Responsible for the Resistance of Local Tomato Accessions to Whitefly (Bemisia tabaci, Gennadius 1889) Hemiptera in Tanzania. *Crops* **2022**, *2*, 445–460. [CrossRef]
15. Acharya, R.; Shrestha, Y.K.; Sharma, S.R.; Lee, K.-Y. Genetic diversity and geographic distribution of Bemisia tabaci species complex in Nepal. *J. Asia Pac. Entomol.* **2020**, *23*, 509–515. [CrossRef]
16. Gill, H.K.; Garg, H.; Gill, A.K.; Gillett-Kaufman, J.L.; Nault, B.A. Onion thrips (Thysanoptera: Thripidae) biology, ecology, and management in onion production systems. *J. Integr. Pest Manag.* **2015**, *6*, pmv006. [CrossRef]
17. Alam, M.; Islam, M.; Haque, M.; Humayun, R.; Khalequzzaman, K.M. Bio-rational management of whitefly (Bemisia tabaci) for suppressing tomato yellow leaf curl virus. *Bangladesh J. Agric. Res.* **2016**, *41*, 583–597. [CrossRef]
18. Ghelani, M.; Kabaria, B.; Ghelani, Y.; Shah, K.; Acharya, M. Biology of whitefly, Bemisia tabaci (Gennadius) on tomato. *J. Entomol. Zool. Stud.* **2020**, *8*, 1596–1599.
19. Ghosh, S.; Ghanim, M. Factors determining transmission of persistent viruses by Bemisia tabaci and emergence of new virus–vector relationships. *Viruses* **2021**, *13*, 1808. [CrossRef]
20. URT. National Sample Census of Agriculture 2019/20. 2021. Available online: https://www.nbs.go.tz/index.php/en/census-surveys/agriculture-statistics (accessed on 15 December 2022).
21. Palacios-Lopez, A.; Christiaensen, L.; Kilic, T. How much of the labor in African agriculture is provided by women? *Food Policy* **2017**, *67*, 52–63. [CrossRef] [PubMed]
22. Kongela, S.M. Gender Equality in Ownership of Agricultural Land in Rural Tanzania: Does Matrilineal Tenure System Matter? *Afric. J. Land Policy Geospat. Sci.* **2020**, *3*, 13–27.
23. Mwatawala, H.W.; Mponji, R.; Sesela, M. Factors Influencing Profitability of Small-Scale Tomato (*Lycopersicon esculentum*) Production in Mvomero District, Tanzania. *Int. J. Progress. Sci. Technol.* **2019**, *14*, 114–121.
24. Masunga, A.W. Assessment of Socio-Economic and Institutional Factors Influencing Tomato Productivity Amongst Smallholder Farmers: A Case Study of Musoma Municipality, Tanzania. Master's Thesis, Sokoine University of Agriculture, Morogoro, Tanzania, 2014. Available online: https://www.suaire.sua.ac.tz/handle/123456789/599 (accessed on 12 September 2022).
25. Anang, B.T.; Zulkarnain, Z.A.; Yusif, S. Production constraints and measures to enhance the competitiveness of the tomato industry in Wenchi Municipal District of Ghana. *Am. J. Exp. Agric.* **2013**, *3*, 824. [CrossRef]
26. Hlouskova, Z.; Prasilova, M. Economic outcomes in relation to farmers' age in the Czech Republic. *Agric. Econ.* **2020**, *66*, 149–159. [CrossRef]
27. Rapsomanikis, G. *The Economic Lives of Smallholder Farmers. An Analysis Based on Household Data from Nine Countries*; Food and Agriculture Organization of the United Nations (FAO): Rome, Italy, 2015.
28. Moranga, L.O.; Otieno, D.J.; Oluoch-Kosura, W. Analysis of Factors Influencing Tomato Farmers' Willingness to Adopt Innovative Timing Approaches for Management of Climate Change Effects in Taita Taveta County, Kenya. Doctoral Dissertation, University of Nairobi, Nairobi, Kenya, 2016.
29. Peng, L.; Zhou, X.; Tan, W.; Liu, J.; Wang, Y. Analysis of dispersed farmers' willingness to grow grain and main influential factors based on the structural equation model. *J. Rural Stud.* **2022**, *93*, 375–385. [CrossRef]
30. Singh, A.K.; Shikha, K.; Shahi, J.P. Hybrids and abiotic stress tolerance in horticultural crops. In *Stress Tolerance in Horticultural Crops*; Elsevier: Amsterdam, The Netherlands, 2021; pp. 33–50.
31. Bala, K.; Sood, A.; Pathania, V.S.; Thakur, S. Effect of plant nutrition in insect pest management: A review: A review. *J. Pharmacogn. Phytochem.* **2018**, *7*, 2737–2742.
32. Vosman, B.; van't Westende, W.P.; Henken, B.; van Eekelen, H.D.; de Vos, R.C.; Voorrips, R.E. Broad spectrum insect resistance and metabolites in close relatives of the cultivated tomato. *Euphytica* **2018**, *214*, 46. [CrossRef]
33. Bihon, W.; Ognakossan, K.E.; Tignegre, J.-B.; Hanson, P.; Ndiaye, K.; Srinivasan, R. Evaluation of Different Tomato (Solanum lycopersicum L.) Entries and Varieties for Performance and Adaptation in Mali, West Africa. *Horticulturae* **2022**, *8*, 579. [CrossRef]
34. Gangwar, R.; Gangwar, C. Lifecycle, distribution, nature of damage and economic importance of whitefly, Bemisia tabaci (Gennadius). *Acta Sci. Agric.* **2018**, *2*, 36–39.
35. Khan, I.; Wan, F. Life history of Bemisia tabaci (Gennadius)(Homoptera: Aleyrodidae) biotype B on tomato and cotton host plants. *J. Entomol. Zool. Stud.* **2015**, *3*, 117–121.
36. Perring, T.M.; Stansly, P.A.; Liu, T.; Smith, H.A.; Andreason, S.A. Whiteflies: Biology, ecology, and management. In *Sustainable Management of Arthropod Pests of Tomato*; Elsevier: Amsterdam, The Netherlands, 2018; pp. 73–110.
37. Nwezeobi, J.; Onyegbule, O.; Nkere, C.; Onyeka, J.; van Brunschot, S.; Seal, S.; Colvin, J. Cassava whitefly species in eastern Nigeria and the threat of vector-borne pandemics from East and Central Africa. *PLoS ONE* **2020**, *15*, e0232616. [CrossRef]
38. Hasanuzzaman, A.T.M.; Islam, M.N.; Zhang, Y.; Zhang, C.-Y.; Liu, T.-X. Leaf morphological characters can be a factor for intra-varietal preference of whitefly Bemisia tabaci (Hemiptera: Aleyrodidae) among eggplant varieties. *PLoS ONE* **2016**, *11*, e0153880. [CrossRef]

39. Fiallo-Olivé, E.; Navas-Castillo, J. Tomato chlorosis virus, an emergent plant virus still expanding its geographical and host ranges. *Mol. Plant Pathol.* **2019**, *20*, 1307–1320. [CrossRef]
40. Dube, J.; Ddamulira, G.; Maphosa, M. Tomato breeding in sub-Saharan Africa-Challenges and opportunities: A review. *Afr. Crop Sci. J.* **2020**, *28*, 131–140.
41. Naveen, N.; Chaubey, R.; Kumar, D.; Rebijith, K.; Rajagopal, R.; Subrahmanyam, B.; Subramanian, S. Insecticide resistance status in the whitefly, Bemisia tabaci genetic groups Asia-I, Asia-II-1 and Asia-II-7 on the Indian subcontinent. *Sci. Rep.* **2017**, *7*, 1–15. [CrossRef]
42. Shankarappa, A.; Marulasidappa, K.C.; Venkataravanappa, V.; Chandrashekar, S. Validation of IPM modules for the management of whitefly, Bemisia tabaci and Mungbean Yellow Mosaic Virus disease in greengram. *J. Entomol. Res. Soc.* **2022**, *24*, 245–255.
43. Patra, B.; Hath, T.K. Insecticide Resistance in Whiteflies Bemisia tabaci (Gennadius): Current Global Status. In *Insecticides—Impact and Benefits of Its Use for Humanity*; Ranz, R.E.R., Ed.; IntechOpen: London, UK, 2022.
44. Castle, S.J.; Merten, P.; Prabhaker, N. Comparative susceptibility of Bemisia tabaci to imidacloprid in field-and laboratory-based bioassays. *Pest Manag. Sci.* **2014**, *70*, 1538–1546. [CrossRef]
45. Abtew, A.; Niassy, S.; Affognon, H.; Subramanian, S.; Kreiter, S.; Garzia, G.T.; Martin, T. Farmers' knowledge and perception of grain legume pests and their management in the Eastern province of Kenya. *Crop Prot.* **2016**, *87*, 90–97. [CrossRef]
46. Damalas, C.A.; Koutroubas, S.D. Farmers' behaviour in pesticide use: A key concept for improving environmental safety. *Curr. Opin. Environ. Sci. Health* **2018**, *4*, 27–30. [CrossRef]
47. Rother, H.-A. Pesticide labels: Protecting liability or health?–Unpacking "misuse" of pesticides. *Curr. Opin. Environ. Sci. Health* **2018**, *4*, 10–15. [CrossRef]
48. Vryzas, Z.; Health. Pesticide fate in soil-sediment-water environment in relation to contamination preventing actions. *Curr. Opin. Environ. Sci. Health* **2018**, *4*, 5–9. [CrossRef]

Disclaimer/Publisher's Note: The statements, opinions and data contained in all publications are solely those of the individual author(s) and contributor(s) and not of MDPI and/or the editor(s). MDPI and/or the editor(s) disclaim responsibility for any injury to people or property resulting from any ideas, methods, instructions or products referred to in the content.

Article

Integrated Control of Scales on Highbush Blueberry in Poland

Małgorzata Tartanus [1,*], Barbara Sobieszek [1], Agnieszka Furmańczyk-Gnyp [2] and Eligio Malusà [1,*]

1. Instytut Ogrodnictwa—PIB, 96-100 Skierniewice, Poland; barbara.sobieszek@inhort.pl
2. Department of Preclinical Veterinary Sciences, Uniwersytet Przyrodniczy w Lublinie, 20-950 Lublin, Poland; agnieszka.furmanczyk@up.lublin.pl
* Correspondence: malgorzata.tartanus@inhort.pl (M.T.); eligio.malusa@inhort.pl (E.M.)

Abstract: In the past decade, the development of highbush blueberry production in Poland has been followed by the occurrence of new pests in the plantations, including scales. Since both the assessment of the populations of natural enemies present in a territory and the knowledge of the scale species present in the crop are crucial for the correct application of IPM strategies, a study was carried out to address these aspects and evaluate the efficacy of several active substances in controlling *Parthenolecanium* spp. in several highbush blueberry plantations. Specimens of adult larvae collected on several plantations were phylogenetically closely linked to two species, *P. corni* and *P. fletcheri*. However, considering the ecology and behavior of these species, it was concluded that the pest population was more likely to belong to *P. corni*. Analyzing the scale parasitoids' community present in the different locations, it emerged that it was quite diversified, including species affecting both the initial and adult biological phases of the scales, with differences also in the population size and diversity, including both general or specialized parasitoids and predators. The different active substances tested in the efficacy trials, which included both synthetic and bio-based compounds, were suitable for controlling the scale infestation. However, the different efficacy observed between them, depending on season and location, could be interpreted taking into consideration the initial level of infestation. It is concluded that applying an IPM strategy that combines agronomical practices with the application of insecticides with different mechanisms of action, attentive to the benefit of protecting natural enemies, can result in satisfactory control of *P. corni* in highbush blueberry plantations.

Keywords: IPM; natural enemies; *Parthenolecanium* spp.; phylogenetic analysis; *Vaccinium corymbosum*

Citation: Tartanus, M.; Sobieszek, B.; Furmańczyk-Gnyp, A.; Malusà, E. Integrated Control of Scales on Highbush Blueberry in Poland. *Horticulturae* **2023**, *9*, 604. https://doi.org/10.3390/horticulturae9050604

Academic Editor: Carmelo Peter Bonsignore

Received: 4 April 2023
Revised: 15 May 2023
Accepted: 17 May 2023
Published: 19 May 2023

Copyright: © 2023 by the authors. Licensee MDPI, Basel, Switzerland. This article is an open access article distributed under the terms and conditions of the Creative Commons Attribution (CC BY) license (https://creativecommons.org/licenses/by/4.0/).

1. Introduction

Highbush blueberry production in Poland has been rapidly developed in the past two decades, allowing the country to become the second largest producer in the EU after Spain, with about 24% of the total EU harvest [1]. This production trend has been followed by the occurrence of "new" pests in the plantations across the country [2,3], including *Parthenolecanium* spp. (Hemiptera: Coccoidea) [4]. Among *Parthenolecanium* spp., several species have been recorded in Poland [5], such as the European fruit lecanium (*P. corni* [Bouchè]), a scale species that is damaging crops either directly or indirectly [6], normally observed on a wide range of fruit host plants (EPPO Global Database, https://gd.eppo.int/taxon/LECACO, accessed on 10 May 2023). The control of *Parthenolecanium* spp. is made difficult by their biological cycle and phenology, requiring that insecticide applications target crawler (mobile instar) emergence to achieve an effective reduction of the scale insect population [7].

Integrated Pest Management (IPM), which is compulsorily applied in the European Union since January 2014, can benefit from the populations of natural enemies present in a territory [8,9]. Important natural enemies of soft scales include predators, particularly members of the Anthribidae and Coccinellidae (Coleoptera) [10], as well as parasitoids, mainly belonging to the hymenopterous family Encyrtidae [11]. Outbreaks of scale insects have been explained by the reduction of natural enemies [12] and eventually as an effect of pesticide application [13].

Scale insects management is complicated by difficulties in their morphological identification [14]. This condition could result in less effective control by using methods or products not fully appropriate for the species present in the crop. However, the recent development of DNA barcoding techniques has also provided useful information for classifying scale insects [15,16]. Recently, specimens of a soft-scale insect were found on *Vaccinium corymbosum* cultivar 'Bluecrop' in Poland, which seemed to be close to *Parthenolecanium corni* (Bouché) based on their morphology but showed significant differences in their life cycle and in their settlement sites.

Therefore, a study was performed to (i) evaluate the efficacy in the control of *Parthenolecanium* spp. under field conditions in several locations and seasons with products having different mechanisms of action; (ii) assess the composition of natural enemies' populations in these orchards; and (iii) determine whether specimens collected in these orchards should be considered congeneric with *P. corni* or whether they formed a separated (undescribed) species.

2. Materials and Methods

2.1. Taxonomic Identification of Scales Sampled from Blueberry Plantations

2.1.1. DNA Extraction, Amplification, and Sequencing

Young adult females of *Parthenolecanium* sp. were collected on *Vaccinium corymbosum* cv. Bluecrop in various locations (Table 1) and stored at $-20\ °C$ until analysis. The DNA extraction was performed using the CTAB method [17], with slight modifications. The DNA concentration was estimated on a 1.5% agarose gel and compared with GeneRuler 100 bp DNA Ladder Plus (Thermo Fisher Scientific, Waltham, MA, USA). Next, the DNA samples were diluted to give a concentration of 20 ng/µL and stored at $-20\ °C$ for downstream analyses.

To study the phylogenetic relationships among specimens, a fragment of DNA containing the mitochondrial cytochrome oxidase subunit I (COI) gene was amplified. The PCR amplifications were performed in a total volume of 25 µL containing: 20 ng/µL of template DNA, 2.5 µL 10 × Buffer Taq (750 mM Tris HCl pH 8.8; 200 mM $(NH_4)_2SO_4$, 0.1% Tween 20) (Thermo Fisher Scientific Waltham, MA, USA), $MgCl_2$ 1500 µM, dNTP mix 800 µM, 0.2 µM of each primer, and 1.0 U Taq polymerase (Fermentas AB, Vilnius, Lithuania). The COI (mtDNA) was amplified with the primer pairs PcoF1 and HCO (Table 1). PCR conditions for COI were set as follows: an initial denaturation at 94 °C for 2 min, followed by 40 cycles of denaturation, annealing, and extension, and a final extension step at 72 °C for 10 min. For the COI amplification, the 35 cycles consisted of 30 s at 95 °C, 50 s at 50 °C, and 1 min at 72 °C. A negative control was used for PCR reactions.

PCR products with an addition of fluorescent dye were separated electrophoretically in a 1.5% agarose gel at 80 V for 1.5 h in 1 × TBE buffer containing 0.01% EtBr and visualized under UV light. After checking and determining the size of the resulting PCR product, the DNA was subjected to purification using an agarose gel of low melting point. The sequencing was conducted by Genomed S.A. (Warsaw, Poland) using the PCR primers together with a Big Dye® Terminator Cycle Sequencing Kit V. 3.1 of Applied Biosystems (Life Technologies, Warsaw, Poland) and separated using a capillary sequencer 3730XL DNA Analyzer.

2.1.2. DNA Sequence Alignment and Phylogenetic Analysis

DNA COI fragments of five *Parthenolecanium* species (*P. corni*, *P. fletcheri*, *P. persicae*, *P. pomeranicum*, and *P. pruinosum*) (Table 2) were included in the sequence alignment to compare genetic diversity with the specimens collected on *V. corymbosum*. Sequences were assembled and edited using BioEdit v. 7.2.5. Related DNA sequences of COI were compared using the BLAST function of GenBank. Multiple sequence alignments were performed with Clustal W using BioEdit v. 7.2.5 [18].

Table 1. Samples of *Parthenolecanium* species used in the molecular analyses.

Scale Insect Species	Host	Locality	Isolate	GenBank No. Sequences
Parthenolecanium corni (Bouché)	No data	Chile	24245	KY085297
	No data	China	S3_499	KP189846
Parthenolecanium fletcheri (Cockerell)	No data	No data	PARFLE	MZ 567176
	Platycladus orientalis	Chungcheongbuk-do, Korea		MK543920
Parthenolecanium persicae (Fabricius)	No data	Australia	WIL	KY362203
	No data	China	S4_079c	KP189853
Parthenolecanium pomeranicum (Kawecki)	*Taxus* sp.	Poland	PAR_POM	MN603162
Parthenolecanium pruinosum (Coquillett)	*Vitis* sp. L.	Australia	Mudgee_2a	KC784924
	Vitis sp. L.	Australia	Gumeracha_b	KC784923
Parthenolecanium sp.	*Vaccinium corymbosum* 'Bluecrop'	Piskórka, Poland	CBr3	ON899817
	Vaccinium corymbosum 'Bluecrop'	Piskórka, Poland	CBr5	ON899818
	Vaccinium corymbosum 'Bluecrop'	Piskórka, Poland	CBr9	ON899819
	Vaccinium corymbosum 'Bluecrop'	Piskórka, Poland	CBr10	ON899820
	Vaccinium corymbosum 'Bluecrop'	Maciejowice, Poland	CBm11	ON899821
	Vaccinium corymbosum 'Bluecrop'	Maciejowice, Poland	CBm13	ON899822
	Vaccinium corymbosum 'Bluecrop'	Maciejowice, Poland	CBm14	ON899823

Table 2. Primers and PCR protocols used.

Gene	Primer	F or R	Primer Sequence 5′ to 3′	References
COI	PcoF1	F	CCTTCAACTAATCATAAAAATATYAG	[19]
	HCO	R	TAAACTTCAGGGTGACCAAAAAATCA	[20]

2.2. Assessment of Parthenolecanium spp. Parasites

Parasitism of *Parthenolecanium* spp. Larvae was determined on sampled shoots (three shoots from four sites in each plantation) that showed the presence of scale females, which were collected during three observation periods (April, May, and June). The healthy females and those with clear parasite damage or parasites inside were separately counted.

Specimens of parasites/parasitoids were obtained by keeping the sampled shoots with the female scales in isolators until the hatched parasites were visible and systematically collected. The specimens were then sent to the Natural History Museum of London for identification.

2.3. Trials of Parthenolecanium sp. Control on Blueberry Plantations

Trials testing different products and strategies for the control of *Parthenolecanium* spp. Were conducted in 2017–2019 on several plantations of highbush blueberry cv. Bluecrop located in four locations in Mazovian Voievodship (Central Poland). The application of control products was carried out during the period of the overwintering larvae's migration.

The experimental field design in each trial consisted of four replications arranged in randomized blocks. Each plot (replication) covered 52.55 m² (three rows of 15 m length, for a total of 45 plants). Tested and reference products having different mechanisms of action (Table 3) were applied with a motorized knapsack sprayer ("Stihl SR 420") with a spray volume of 750 L/ha.

Table 3. List of active substances utilized in the field trials for the control of highbush blueberry scales.

Active Substance	Product	Mechanism of Action According to IRAC	Activity on the Pest
Camelina oil (*Camelina sativa* (L.) Crantz)	Emulpar 940 EC	N/A	Mechanical action and suffocation
Spinosad	SpinTor 240 EC	Nicotinic acetylcholine receptor (nAChR) allosteric modulators (Site I)	Contact and stomach poisoning, and ovicidal
Silicon polymers	Insect Control	N/A	Mechanical action and suffocation.
Silicon polymers	Siltac EC	N/A	Mechanical action, suffocation
Acetamipryd	Mospilan 20 SP Stonkat 20 SP	Nicotinic acetylcholine receptor (nAChR) competitive modulators	Contact and stomach poisoning
Flonicamid	Tepekki 50 WG	Chordotonal organ and Modulators—undefined target sites	Systemic
Spirotetramat	Movento 100 SC	Inhibitors of acetyl CoA carboxylase	Systemic

The scale population density was estimated just before the treatment and approximately 2–3 weeks after spraying. The number of alive scale larvae was counted under a stereoscopic microscope on three stems (each 30 cm long) taken randomly from each plot/replication on each counting date.

2.4. Statistical Analysis

2.4.1. Phylogenetic Analysis

The phylogenetic analysis was performed with MEGA11 software [21], and the evolutionary history was inferred using the Maximum Likelihood method based on the Tamura-Nei model [22]. The bootstrap consensus tree was inferred from 1000 replicates [23]. Branches corresponding to partitions reproduced in less than 50% of bootstrap replicates collapsed. The initial tree(s) for the heuristic search were obtained automatically by applying the Neighbor-Join and BioNJ algorithms to a matrix of pairwise distances estimated using the Tamura-Nei model and then selecting the topology with the superior log likelihood value. This analysis involved 16 nucleotide sequences. Codon positions included were 1st, 2nd, 3rd, and noncoding. All positions containing gaps and missing data were discarded. There were a total of 378 positions in the final dataset. The nucleotide sequences were compared with sequences collected in the NCBI GenBank databases using BLAST software [24] (http://www.ncbi.nlm.nih.gov/BLAST/, accessed on 15 January 2023).

The Bayesian inference analysis was carried out using the Markov Chain Monte Carlo (MCMC) algorithm with the mrbayes program ver. 3.2.2 [25], using Biolinux 8.0 as OS, and applying the default parameters.

2.4.2. Field Data Analysis

The data from field trials were analyzed by ANOVA performed on values transformed by log(x) + 1 in order to assure a normal distribution. The significance of differences between means was assessed with the Tukey multiple range test at $p \leq 0.05$ using the package Statistica v.6.1. The value of actual mortality was calculated according to Abbott's formula [26].

3. Results

3.1. Taxonomic Identification of Highbush Blueberry Scale Specimens

A phylogenetic tree was constructed based on the sequence of the COI gene fragments obtained from the *Parthenolecanium* sp. specimens collected in the study and the sequences of the genus *Parthenolecanium* deposited in GenBank (Figure 1). Two main clades can be distinguished in the trees obtained with both bootstrap and Bayesian analyses. The first clade is formed by the species *P. prunoisum*, while the second is formed by *P. pomeraniucum*, *P. persicae*, *P. fletcheri*, and *P. corni*, including the specimens from *V. corymbosum* as well. Two branches can be distinguished within the second clade, one of which contains *P. pomeranicum*. The specimens collected from blueberry plants were located on the second branch, positioned close to *P. fletcheri* and *P. corni*, indicating a close relationship between these taxa.

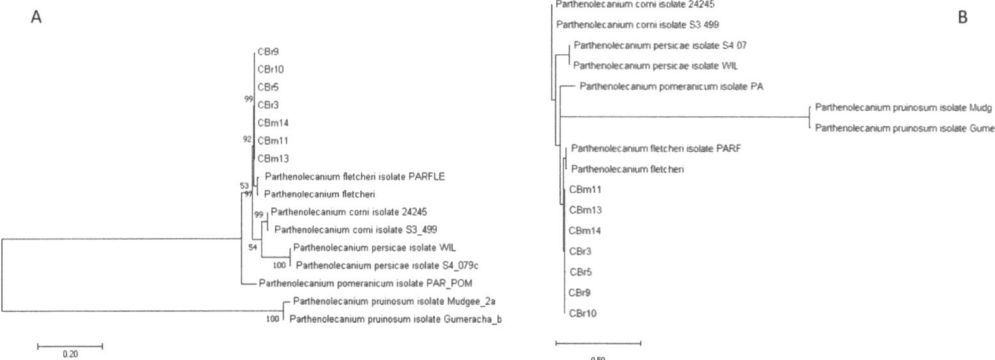

Figure 1. Dendrograms showing the phylogenetic relationship of the adult females scale specimens collected on *V. corymbosum* with isolates of different *Parthenolecanium* species based on the analysis of molecular data by the maximum likelihood method (**A**) and Bayesian inference analysis (**B**). The percentage of replicate trees in which the associated taxa clustered together in the bootstrap test is shown below the branches.

3.2. Level of Parasitism of Parthenolecanium sp. under Field Conditions

The number of larvae found during the assessment was, on average, about 240 per shoot. The dynamic of the population and the percentage of parasitized larvae varied depending on the location (Figure 2). However, a clear difference in the dynamic was observed comparing the plantation in Piskórka with the other three: the increase (more than doubled) in the population size in Piskórka from April to May was paralleled by an increase of the parasitized larvae, while in the other two locations the population size was either steady (Rokotów) or decreased (Prazmów) in a similar pattern for healthy or parasitized larvae (Figure 2). The Piskórka plantation was also characterized by having almost 20% parasitized larvae during the whole season, ranging from 15% up to six times that of the other three plantations throughout the whole period of assessment (Figure 3).

3.3. Identification of Parthenolecanium spp. Parasites

The number of parasitic species or genera found in the four plantations ranged from four to nine (Table 4). The plantation located in Prażmów was characterized by the highest biodiversity in this respect, also including a predator (*Anthribidae* spp.) not found in other locations. On the other hand, three species (*Blastothrix brittanica*, *Coccophagus lycimnia*, and *Encyrtus infelix*) were common to all sites, and two (*Syrphophagus taeniatus* and *Metaphycus insidiosus*) were found in three sites. One genus, *Scelionidae* spp., was found only in one location (Jakubów).

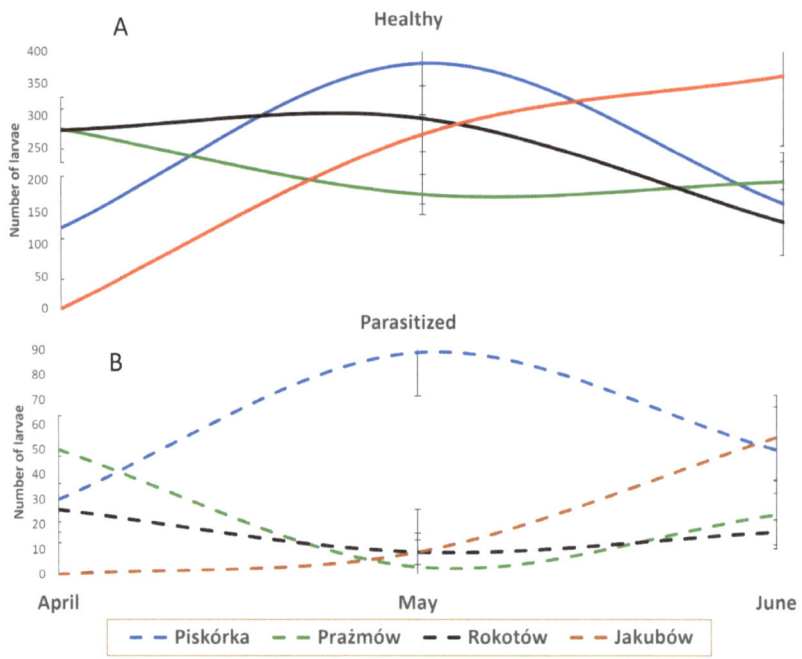

Figure 2. Dynamic impact of natural parasites on the infestation of highbush blueberry plantations by *Parthenolecanium* spp.: number of healthy (**A**) and parasitized (**B**) larvae. Bars represent SD, n = 12.

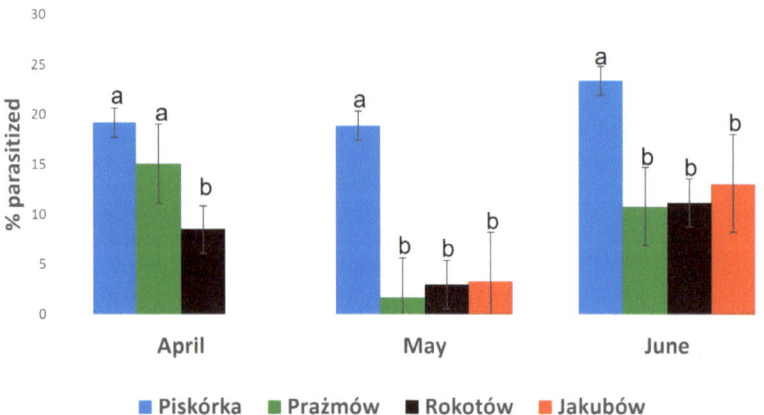

Figure 3. Percent of parasitized *Parthenolecanium* larvae in the sampled population. Mean ± SD, n = 12. Different letters at each time point represent the statistical difference for $p \leq 0.05$.

3.4. Control of Parthenolecanium sp. on Blueberry Plantations

The infestation by the scales varied depending on the season and the location, averaging from a few individuals to some hundreds (Tables 5–7). However, interestingly, the same location, Rokotów, recorded both the lowest (on average 2.5 individuals per shoot) and the highest (on average about 190 individuals per shoot) infestations in 2018 and 2019, respectively.

Table 4. List of the parasites/parasitoids identified from larvae of *Parthenolecanium* spp. collected from different highbush blueberry plantations.

Parasite Species or Genera	Location of the Plantation			
	Jakubów	Piskórka	Prażmów	Rokotów
Blastothrix spp. including *B. brittanica* (Girault, 1917)	X	X	X *	X
Coccophagus lycimnia (Walker, 1839)	X	X	X	X
Encyrtus infelix (Embleton, 1902)	X	X	X	X
Metaphycus spp. including *M. insidiosus* (Mercet, 1921)		X	X *	X
Syrphophagus spp. including *S. taeniatus* (Förster, 1861)		X *	X	X
Anthribidae spp.			X	
Scelionidae spp.	X			
Total	4	5	6	5

* Location where specimens classified both at species and genus levels were found.

Table 5. Effect of different products on the control of *Parthenolecanium* sp. on different blueberry plantations in 2017. Mean ± SD, n = 12. Different letters in columns represent statistical differences for $p \leq 0.05$.

Treatment	Dose Applied	Living Larvae per Shoot (n)	
		Piskórka	Maciejowice
Before treatment	-	9.3 ± 26.9	10.0 ± 8.5
Control	-	7.9 ± 2.2 b	34.3 ± 16.6 b
Spinosad	0.4 L/ha	5.0 ± 6.7 b	17.3 ± 17.3 a
Silicon polymers	0.2%	1.7 ± 1.1 ab	11.0 ± 15.4 a
Spirotetramat	0.75 L/ha	1.0 ± 2.1 a	12.7 ± 11.4 a

Table 6. Effect of different products on the control of *Parthenolecanium* sp. on different blueberry plantations in 2018. Mean ± SD, n = 12. Different letters in columns represent statistical differences for $p \leq 0.05$.

Treatment	Dose Applied	Living Larvae per Shoot (n)			
		Rokotów	Piskórka I	Piskórka II (Tunel)	Prażmów
Before treatment	-	2.5 ± 3.0	19.5 ± 18.2	20.7 ± 14.2	79.2 ± 33.2
Control	-	1.4 ± 1.3 b	9.3 ± 5.1 b	4.7 ± 3.1 b	15.2 ±6.6 b
Camelina oil	1.2%	0.6 ± 0.8 ab	1.0 ± 2.1 a	1.2 ±3.5 a	6.6 ± 13.5 ab
Acetamipryd	0.2 kg/ha	0.5 ± 1.5 ab	1.3 ± 6.4 a	1.3 ± 5.8 a	8.7 ±20.9 ab
Flonicamid	0.14 kg/ha	1.6 ± 4.2 b	1.4 ± 1.5 a	1.5 ± 7.8 a	4.7 ± 9.3 ab
Spirotetramat	0.75 L/ha	0.0 ± 0.0 a	0.9 ± 3.7 a	0.7 ± 2.9 a	2.1 ± 6.8 a

Table 7. Effect of different products on the control of *Parthenolecanium* sp. on different blueberry plantations in 2019. Mean ± SD, n = 12. Different letters in columns represent statistical differences for $p \leq 0.05$.

Treatment	Dose Applied	Living Larvae per Shoot (n)	
		Piskórka	Rokotów
Before treatment	-	187.5 ± 36.7	336.0 ± 32.0
Control	-	15.5 ± 7.4 c	248.2 ± 30.4 c
Spinosad	0.4 L/ha	6.1 ± 2.9 bc	5.5 ± 9.1 a
Silicon polymers	0.15%	4.8 ± 7.8 bc	17.3 ± 20.6 b
Acetamipryd	0.2 L/ha	1.2 ± 2.9 a	8.0 ± 12.5 ab
Flonicamid	0.14 kg/ha	6.7 ± 5.4 bc	12.9 ± 9.1 ab

The different products applied, expressing different mechanisms of action, including some allowed in organic farming (i.e., camelina oil and spinosad), were in general effective in reducing the number of living larvae (Figure 4), but their efficacy was also influenced by the location, season, and initial level of infestation (Tables 5–7). For example, in the case of the synthetic active substances, during the three years of trials at the Piskórka plantation, only spirotetramat consistently reduced the number of living larvae significantly compared to the control, reaching an efficacy of 85–100%. Flonicamid or acetamiprid instead had a more variable effect, both when comparing the same location across the years (Piskórka) or confronting different locations (Piskórka and Rokotów) or the initial level of infestation (Tables 5–7), resulting in a lower efficacy than spirotetramet (Figure 4). The products allowed in organic farming (camelina oil and spinosad), as well as those based on silicon polymers, also had a variable effect, influenced by the same factors as the synthetic substances (Tables 5–7). However, in some cases, their efficacy was similar to the synthetic molecules, even for the protected crops, thus showing good control potential.

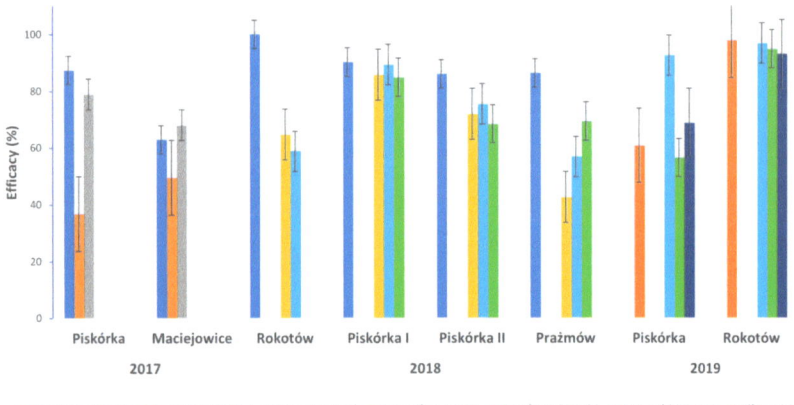

Figure 4. Efficacy rate of the different active substances in controlling larvae of *Parthenolecanium* spp. on different blueberry plantations in 2017–2019. Mean ± SD, n = 12.

4. Discussion

4.1. Phylogenetic Identification of the Specimens

The specimens collected from the highbush blueberry orchards were grouped together, thus indicating a common genetic relationship, even though they were collected from orchards located in different regions. All specimens were phylogenetically closely linked to two other species, *P. corni* and *P. fletcheri*, which practically did not show genetic distance between them and the collected specimens. The dendrograms visually suggested a closer relationship between the specimens and *P. fletcheri* compared to the *P. corni* samples analyzed, but the genetic distance resulted in similar results. Both species have been recorded in Poland [5]. However, *P. fletcheri*, even though it was recorded for the first time in Poland in 1935, is still considered an alien species [27]. *P. corni* and *P. fletcheri* are morphologically very similar and share high variability in many morphological characters that are normally used in classification [28]. Moreover, considering that *P. corni* is a widespread species normally found on fruit trees, while *P. fletcheri* tends to be found mainly on a few forest tree species, it is more likely that the specimens analyzed belonged to the former species.

4.2. Ecology of Parthenolecanium sp. Parasites

Genera and species found to parasitize *Parthenolecanium* spp. in Polish highbush blueberry plantations are known natural enemies, either as parasitoids (*Blastothrix*, *Coccophagus*, *Encyrtus*, and *Metaphycus*) or predators (*Anthribus*) [29–31]. *Coccophagus lycimnia* and *Blastothrix* species are cosmopolitan, generalist parasitoid taxa that attack many scale species [30,32], so it was not surprising to find them in all the highbush blueberry plantations. Moreover, *Blastothrix longipennis*, *Metaphycus insidiosus*, and *Coccophagus lycimnia* were reported to be common and important parasitoids of *P. corni* in Poland [33]. Interestingly, the number of parasitoid genera found in association with *Parthenolecanium* spp. under Polish conditions was similar to that found in a fruit-producing region of Bulgaria [34]. Under those conditions, *C. lycimnia* and *B. confusa* were found to have the greatest importance in regulating the population density of the pest, as they have different targets: *C. lycimnia* parasitizing overwintering larvae and *B. confusa* parasitizing adult females.

Analyzing the parasitoid community attacking *P. corni* at different stages of its development, it emerged that it was quite diversified in its range, affecting both initial and adult biological phases, but also in its occurrence in different locations, as more species were present in Prażmów compared to Jakubów or Rokotów. Moreover, Prażmów was also the only location where a predator of the *Anthribidae* genera was observed. The presence of natural coniferous woods surrounding the plantation at Prażmów could suggest a possible explanation for this finding, as scales are frequently found on such and other forest plants under Polish conditions [35,36], thus giving the possibility of developing a complex population of natural enemies. This would also support the only finding in this location of *Anthribidae* spp., a predator species, as ladybirds can be an effective natural enemy of coniferous scales [37]. *Anthribus nebulosus* is an effective natural enemy of scale insects, including *P. corni*, as its larvae can act as parasitoids of adult scale individuals, while adults are able to act as predators in all stages of their hosts [9,38,39]. The efficacy of *A. nebulosus* in reducing the populations of *P. corni* was about 22% in Serbia [40].

The dynamic of the healthy or parasitized scale populations recorded in Jakubów, a location densely filled with fruit orchards, could be considered a classic example of the relation between pest and parasite populations [41], with the latter mainly formed by generalistic species, including scelionids. Most scelionids are solitary, and all are egg parasitoids that utilize the eggs of a wide variety of insects [42]. However, species of this family have been used successfully in classical biological control programs [43]. The recent release of *Trissolcus japonicus* to control *Halyomorpha halys* [44] is also an example of the potential biocontrol from using specific parasitoids of this family for *Partenolecanium* spp. control. Even though the level of parasitization was much higher in Piskórka, a site characterized by a natural environment rich in coniferous woods, compared to the other plantations in all three sampling periods, the parasitization rate found in the highbush blueberry plantations was in general low compared to other reports from other fruit growing areas [34].

4.3. Is It Possible to Develop a Strategy for the Integrated Control of Parthenolecanium sp. in Highbush Blueberry Orchards with Low Environmental Impact?

The different active substances tested in the trials proved suitable for controlling the scale infestation. However, the different efficacy observed between them should be interpreted taking into consideration the initial level of infestation. The high efficiency of Spinosad in the Rokotów 2019 trial was obtained when the untreated control had a high infestation level, almost unchanged from the assessment before the treatments. The same substance was less effective when the population in control plots was drastically reduced during the assessment period (Piskórka in 2019), likely as a result of the impact of natural enemies. In the latter case, only acetamiprid, a chloropyridinyl neonicotinoid, was highly effective. However, even though it was consistent in its efficacy, acetamipirid was sometimes less effective than the other synthetic substances (e.g., Rokotów and Prazmów in 2018). The use of acetamiprid for the control of the 1st instar crawler induced almost 97% mortality 21 days after the first treatment [45].

Spinosad, a bio-insecticide derived from the soil actinomycete *Saccharopolyspora spinosa*, is considered a valuable bioactive substance to control several pests [46]. It was found to be effective in controlling *P. corni* in grapes, but only at the beginning of the infestation [47]. Even though spinosad is classified as a substance with reduced environmental and toxicological risk [48,49], its toxicity on several hymenopteran parasitoids has been reviewed recently [50], and it was shown to cause 100% mortality on *C. lycimnia*, a specific parasitoid of *P. corni*, just 24 h after the treatment [51].

The silicon polymer showed intermediate efficacy in all trials. However, when the infestation in the control was high (Maciejowice in 2017 or Rokotów in 2019), the reduction of the number of larvae was significant. When applied to plants, the silicone polymer spreads on the treated surface, creating a three-dimensional grid structure with sticky properties that blocks the insect's physical functions [52]. The efficacy of this kind of substance can thus depend on the application method and on the environmental conditions, as was also previously shown [4].

Camelina oil, characterized by a high content of poly-unsaturated and saturated fatty acids [53], was also more effective with a high infestation rate, including when the crop was grown under protected conditions. The mode of action for many oils is suffocation and water loss [54], and therefore they are unlikely to induce resistance in insect populations [55]. A canola oil treatment resulted in very high efficacy with concentrated sprays (90% with 20 L/ha and 99% with 30 L/ha), similar to that obtained with a tank of the same oil mix (1 L/ha) with chlorpyrifos-methyl [56]. Paraffin oil reduced the number of *P. corni* larvae by an average of about 80% in trials under different pedo-climatic conditions [57]. Other non-synthetic substances suitable for scale control under organic management of highbush blueberries were reported to have different efficacy. A mixture of vegetal amino acids and fatty acids or treatment with *Quassia amara* extract 50% and potassium soap showed insufficient efficacy [56]. On the other hand, a product based on polysaccharides and also containing chitosan showed a satisfactory level of efficacy [4]. The combined and alternated use of this kind of substance for the control of scales on highbush blueberries could thus represent a good strategy to reduce the environmental impact and improve the overall control efficacy.

The difficulty in controlling scales can be assessed considering the results from the treatments with synthetic active substances. The consistent and high efficacy shown by spirotetramat across locations and seasons, confirming previous results [4], can allow us to consider this product as a reference standard. However, spirotetramat efficacy changed significantly across seasons in a 2-year trial in Slovenia, from 98% in the first year to 70% in the second [58]. The efficacy of flonicamid, a selective systemic pesticide that interferes with the alimentary behavior of several insect species [59,60], was somehow inconsistent throughout the seasons and locations in the present study; its efficacy was significant only in two out of six trials where it was tested. However, as it was shown to have a small impact on parasitoids of other Coccidae species, causing the lowest reduction in their parasitism rate [61], its application and efficacy should be further verified.

Scale populations could remain below the economic threshold thanks to natural biological control, and it is believed that this has been the case for highbush blueberry plantations in Poland in the past, as no major need for their control was raised by producers. However, it could be argued that the need to assure protection against *D. suzukii* in recent years also in Poland, with increased use of pesticides, might have disrupted the natural control in highbush blueberry orchards, similarly to other cases [47,62], requiring the performance of some control measures against scales.

The presence of several genera and species of parasitoids in the highbush blueberry plantations concerned by the research allows us to envision their contribution to *P. corni* scale control even when pesticides are applied. Indeed, while insecticides are generally toxic to adult scelionids, it appears that preimaginal parasitoids within host eggs could escape high mortality in the field even when formulations based on insect growth regulators are applied [63]. Nevertheless, in addition to direct mortality, pesticides sublethal effects on beneficial insects' physiology and behavior can also modify the population dynamics and biological control potential of scale parasitoids [11], thus requiring a careful selection of substances to be applied.

Control of scales is a major challenge because adult females and eggs are protected from pesticides, with only the first-instar nymphs being susceptible due to their mobility, making the timing of pesticide applications targeting them critical to achieving significant efficacy. A good integrated strategy to manage scales includes dormant pruning of old, weak canes and scale-infested wood, which prevents increasing their population density. Applying a complex strategy, combining the application of several synthetic insecticides with different MoAs and timings, allowed us to achieve an efficacy above 90% [64]. A good control was also possible by introducing a control based on non-synthetic substances (this report; [4]). The importance of *P. corni* control is also derived from its function as a vector of several viruses [65,66]. Recently, it was shown that larval stages and adult females of *P. corni* feeding on highbush blueberry plants infected by the blueberry red ringspot virus carried the virus, even though it was not proven they were able to transmit it [67].

5. Conclusions

The substances tested for the control of *P. corni* scales on highbush blueberry during several years and in different locations were suitable, but the level of efficacy depended on various factors and should be interpreted taking also into consideration the initial level of infestation. It is argued that the infestation level could have been affected by the population composition of natural enemies, which was diversified at the studied sites. Therefore, applying an IPM strategy that combines agronomical practices with the application of insecticides with different mechanisms of action, attentive to the benefit of protecting natural enemies, can result in satisfactory control of *P. corni* in highbush blueberry plantations. Even though the scales present in the highbush blueberry plantations presented some uncommon morphological characteristics, the comparison based on COI sequences of their DNA with those of other accessions from *P. corni* or *P. fletcheri* did not show significant differences.

Author Contributions: Conceptualization, M.T.; methodology, M.T. for field trials and A.F.-G. for phylogenetic analysis; laboratory analyses, B.S. for field samples, A.F.-G. for phylogenetic analysis; investigation, M.T. and E.M.; data curation, M.T.; writing—original draft preparation, E.M. and M.T.; writing—review and editing, all authors. funding acquisition, M.T. All authors have read and agreed to the published version of the manuscript.

Funding: This research was funded by the Polish Ministry of Education and Sciences (grant number ZORpSz/3/2016–d. 2.2.2); "Monitoring inwazyjnych owadów i roztoczy zagrażających uprawom sadowniczym oraz opracowanie biologicznych podstaw ich zwalczania".

Data Availability Statement: The data presented in this study are available in the article.

Acknowledgments: Christina Fisher and Max Barclay from the Natural History Museum of London is acknowledged for the identification of the scale parasites. Barbara Łagowska i Katarzyna Golan from Uniwersytet Przyrodniczy w Lublinie for the advices on phylogenetic analyses. The support from the farmers during the field trials is also acknowledged.

Conflicts of Interest: The authors declare no conflict of interest.

References

1. International Blueberry Organization. Global State of the Blueberry Industry Report 2022. Available online: https://www.internationalblueberry.org/2022-report/ (accessed on 28 March 2023).
2. Łabanowska, B.H.; Piotrowski, W. The spotted wing drosophila *Drosophila suzukii* (Matsumura, 1931)—Monitoring and first records in Poland. *J. Hortic. Res.* **2015**, *23*, 49–57. [CrossRef]
3. Kalinowska, E.; Paduch-Cichal, E.; Chodorska, M.; Sala-Rejczak, K. First report of Blueberry red ringspot virus in highbush blueberry in Poland. *J. Plant Pathol.* **2011**, *93*, Supplement pp. S4.73.
4. Tartanus, M.; Malusá, E.; Sas, D.; Łabanowska, B. Integrated control of Lecanium scale (*Parthenolecanium* spp.) on highbush blueberry in open field and protected crops. *J. Plant Prot. Res.* **2018**, *58*, 297–303. [CrossRef]
5. Łagowska, B. Czerwce (Coccoidea), Zabielicowate (Ortheziidae), Czerwcowate (Margarodidae), Czerwce mączyste (Pseudococcidae), Pilśnikowate (Eriococcidae), Kermesowate (Kermesidae), Miłkowate (Cerococcidae), Miseczniкowate (Coccidae), Gwiazdosze (Asterolecaniidae), Tarczniki (Diaspididae). In *Fauna Polski—Charakterystyka i wykaz gatunków*; Bogdanowicz, W., Chudzicka, E., Pilipiuk, I., Skibińska, E., Eds.; Muzeum and Instytut Zoologii PAN: Warszawa, Poland, 2004; pp. 266–269.
6. Vranjic, J.A. Effects on host plant. In *Soft Scale Insects: Their Biology, Natural Enemies and Control*; Ben-Dov, Y., Hodgson, C.J., Eds.; Elsevier Science B.V.: Amsterdam, The Netherlands, 1997; Volume 7A, pp. 323–336.
7. Robayo Camacho, E.; Chong, J.-H. General biology and current management approaches of soft scale pests (Hemiptera: Coccidae). *J. Integr. Pest. Manag.* **2015**, *6*, 1–22.
8. Stark, J.D.; Vargas, R.; Banks, J.E. Incorporating ecologically relevant measures of pesticide effect for estimating the compatibility of pesticides and biocontrol agents. *J. Econ. Entomol.* **2007**, *100*, 1027–1032. [CrossRef]
9. Trdan, S.; Laznik, Ž.; Bohinc, T. Thirty years of research and professional work in the field of biological control (predators, parasitoids, entomopathogenic and parasitic nematodes) in Slovenia: A review. *Appl. Sci.* **2020**, *10*, 7468. [CrossRef]
10. Ponsonby, D.J.; Copland, M.J.W. Coccinellidae and other Coleoptera. In *Soft Scale Insects: Their Biology, Natural Enemies and Control*; Ben-Dov, Y., Hodgson, C.J., Eds.; Elsevier Science B.V.: Amsterdam, The Netherlands, 1997; Volume 7A, pp. 29–60.
11. Kapranas, A.; Tena, A. Encyrtid Parasitoids of Soft Scale Insects: Biology, Behavior, and Their Use in Biological Control. *Ann. Rev. Entomol.* **2015**, *60*, 195–211. [CrossRef]
12. Price, P.W.; Denno, R.F.; Eubanks, M.D.; Finke, D.L.; Kaplan, I. Insect Ecology: Behavior, Populations and Communities. Cambridge University Press: New York, NY, USA, 2011.
13. Raupp, M.J.; Holmes, J.J.; Sadof, C.; Shrewsbury, P.; Davidson, J.A. Effects of cover sprays and residual pesticides on scale insects and natural enemies in urban forests. *J. Arboric.* **2001**, *27*, 203–214. [CrossRef]
14. Gullan, P.J.; Cook, L.G. Phylogeny and higher classification of the scale insects (Hemiptera: Sternorrhyncha: Coccoidea). *Zootaxa* **2007**, *425*, 413–425. [CrossRef]
15. Andersen, J.C.; Wu, J.; Gruwell, M.E.; Gwiazdowski, R.; Santana, S.H.; Feliciano, N.H.; Morse, G.E.; Normark, B.B. A phylogenetic analysis of armored scale insects (Hemiptera: Diaspididae), based upon nuclear, mitochondrial, and endosymbiont gene sequences. *Mol. Phylogenet. Evol.* **2010**, *57*, 992–1003. [CrossRef]
16. Wang, X.-B.; Deng, J.; Zhang, J.-T.; Zhou, Q.-S.; Zhang, Y.-Z.; Wu, S.-A. DNA barcoding of common soft scales (Hemiptera: Coccoidea: Coccidae) in China. *Bull. Entomol. Res.* **2015**, *105*, 1–10. [CrossRef] [PubMed]
17. Doyle, J.J.; Doyle, J.L. A rapid DNA isolation procedure for small quantities of fresh leaf tissue. *Phytochem Bull.* **1987**, *19*, 11–15.
18. Hall, T.A. BioEdit: A User-Friendly Biological Sequence Alignment Editor and Analysis Program for Windows 95/98/NT. *Nucleic Acids Symp. Ser.* **1999**, *41*, 95–98.
19. Park, D.S.; Suh, S.J.; Oh, H.W.; Hebert, P.D.N. Recovery of the mitochondrial COI barcode region in diverse Hexapoda through tRNA-based primers. *BMC Genom.* **2010**, *11*, 423. [CrossRef] [PubMed]
20. Folmer, O.; Black, M.; Hoeh, W.; Lutz, R.; Vrijenhoek, R. DNA primers for amplification of mitochondrial cytochrome c oxidase subunit I from diverse metazoan invertebrates. *Mol. Mar. Biol. Biotechnol.* **1994**, *3*, 294–299.
21. Tamura, K.; Stecher, G.; Kumar, S. MEGA 11: Molecular Evolutionary Genetics Analysis Version 11. *Mol. Biol. Evol.* **2021**, *38*, 3022–3027. [CrossRef]
22. Tamura, K.; Nei, M. Estimation of the number of nucleotide substitutions in the control region of mitochondrial DNA in humans and chimpanzees. *Mol. Biol. Evol.* **1993**, *10*, 512–526.
23. Felsenstein, J. Confidence limits on phylogenies: An approach using the bootstrap. *Evolution* **1985**, *39*, 783–791. [CrossRef]
24. Zhang, Z.; Schwartz, S.; Wagner, L.; Miller, W. A greedy algorithm for aligning DNA sequences. *J. Comput. Biol.* **2000**, *7*, 203–214. [CrossRef]

25. Ronquist, F.M.; Teslenko, P.; van der Mark, D.; Ayres, A.; Darling, S.; Höhna, B.; Larget, L.; Liu, M.; Suchard, A.; Huelsenbeck, J.P. MrBayes 3.2: Efficient Bayesian phylogenetic inference and model choice across a large model space. *Syst. Biol.* **2012**, *61*, 539–542. [CrossRef]
26. Abbott, W.S. A Method of Computing the Effectiveness of an Insecticide. *J. Econ. Entomol.* **1925**, *18*, 265–267. [CrossRef]
27. Łagowska, B.; Golan, K.; Kot, I.; Kmieć, K.; Górska-Drabik, E.; Goliszek, K. Alien and invasive scale insect species in Poland and their threat to native plants. *Bull. Insectology* **2015**, *68*, 13–22.
28. Stepaniuk, K.; Łagowska, B. Number and arrangement variation of submarginal tubercles in adult females *Parthenolecanium corni* group (Hemiptera, Coccidae) and its value as a taxonomic character. *Pol. J. Entomol.* **2006**, *75*, 293–301.
29. Schultz, P.B. Natural enemies of oak lecanium (Homoptera: Coccidae) in eastern Virginia. *Environ. Entomol.* **1984**, *13*, 1515–1518. [CrossRef]
30. Noyes, J.S. Universal Chalcidoidea Database; World Wide Web Electronic Publication, UK: 2019. Available online: http://www.nhm.ac.uk/chalcidoids (accessed on 10 March 2023).
31. Meineke, E.K.; Dunn, R.R.; Frank, S.D. Early pest development and loss of biological control are associated with urban warming. *Biol. Lett.* **2014**, *10*, 20140586. [CrossRef] [PubMed]
32. Robayo Camacho, E.; Chong, J.-H.; Braman, S.K.; Frank, S.D.; Schultz, P.B. Natural enemy communities and biological control of *Parthenolecanium* spp. (Hemiptera: Coccidae) in the Southeastern United States. *J. Econ. Entomol.* **2018**, *111*, 1558–1568. [CrossRef]
33. Moglan, I. Complexes parasitaires de quelques espèces de coccides (Homoptera, Coccidae) en Roumanie. *Entomol. Rom.* **2007**, *12*, 267–275.
34. Arnaoudov, V.; Olszak, R.; Kutinkova, H. Natural enemies of plum brown scale *Parthenolecanium corni* Bouché (Homoptera: Coccidae) in plum orchards in the region of Plovdiv. *IOBC/Wprs Bull.* **2006**, *29*, 105–109.
35. Koteja, J. Notes on the Polish scale insect fauna (Homoptera, Coccoidea). IV. *Polskie Pismo Entomol.* **1972**, *42*, 565–571.
36. Goliszek, K.; Łagowska, B.; Golan, K. Scale insects (Hemiptera, Sternorrhyncha, Coccoidea) on ornamental plants in the field in Poland. *Acta Sci. Pol. Hortorum Cultus* **2011**, *10*, 75–84.
37. Graora, D.; Spasić, R.; Mihajlović, L. Bionomy of spruce bud scale, *Physokermes piceae* (Schrank) (Hemiptera: Coccidae) in the Belgrade area, Serbia. *Arch. Biol. Sci.* **2012**, *64*, 337–343. [CrossRef]
38. Kosztarab, M.; Kozar, F. Introduction of *Anthribus nebulosus* (Coleoptera: Anthribidae) in Virginia for control of scale insects: A review. *Virginia J. Sci.* **1983**, *34*, 223–236.
39. Kosztarab, M.; Kozár, F. *Scale Insects of Central Europe*; Series Entomologica; Akademiai Kiado: Budapest, Hungary, 1988; Volume 41, 456p.
40. Dervisevic, M.; Graora, D. The life cycle and efficacy of *Anthribus nebulosus* Forster in reducing soft scale populations in Belgrade *Fresenius Environ. Bull.* **2019**, *28*, 1981–1985.
41. Kidd, N.A.C.; Jervis, M.A. Population Dynamics. In *Insects as Natural Enemies*; Jervis, M.A., Ed.; Springer: Dordrecht, Germany, 2007. [CrossRef]
42. Masner, L. Revisionary notes and keys to world genera of Scelionidae (Hymenoptera: Proctotrupoidea). *Mem. Ent. Soc. Canada* **1976**, *108*, 1–87. [CrossRef]
43. Greathead, D.J. Opportunities for biological control of insect pests in tropical Africa. *Rev. Zool. Afr.* **1986**, *100*, 85–96.
44. Falagiarda, M.; Carnio, V.; Chiesa, S.G.; Pignalosa, A.; Anfora, G.; Angeli, G.; Ioriatti, C.; Mazzoni, V.; Schmidt, S.; Zapponi, L. Factors influencing short-term parasitoid establishment and efficacy for the biological control of *Halyomorpha halys* with the samurai wasp *Trissolcus japonicus*. *Pest Manag. Sci.* **2023**. [CrossRef]
45. Seok-Min, L.; Bu-Keun, C.; Dong-Wan, K.; Kyung-Mi, P.; In-Young, H.; Jin-Hyeuk, K.; Heung-Su, L. Seasonal development and control of *Parthenolecanium corni* in blueberry Shrubs. *Korean J. App. Entomol.* **2021**, *60*, 403–415.
46. Bacci, L.; Lupi, D.; Savoldelli, S.; Rossaro, B. A review of Spinosyns, a derivative of biological acting substances as a class of insecticides with a broad range of action against many insect pests. *J. Entamol. Acarol. Res.* **2016**, *48*, 40–52. [CrossRef]
47. Duso, C.; Pozzebon, A.; Lorenzon, M.; Fornasiero, D.; Tirello, P.; Simoni, S.; Bagnoli, B. The Impact of Microbial and Botanical Insecticides on Grape Berry Moths and Their Effects on Secondary Pests and Beneficials. *Agronomy* **2022**, *12*, 217. [CrossRef]
48. Williams, T.; Valle, J.; Viñuela, E. Is the naturally derived insecticide Spinosad®compatible with insect natural enemies? *Biocontrol Sci. Technol.* **2003**, *13*, 459–475. [CrossRef]
49. Santana, V.; Santos, V.; Barbosa Pereira, B. Properties, toxicity and current applications of the biolarvicide spinosad. *J. Toxicol. Environ. Health Part B* **2020**, *23*, 13–26. [CrossRef]
50. Biondi, A.; Mommaerts, V.; Smagghe, G.; Viñuela, E.; Zappalà, L.; Desneux, N. The non-target impact of spinosyns on beneficial arthropods. *Pest Manag. Sci.* **2012**, *68*, 1523–1536. [CrossRef] [PubMed]
51. Suma, P.; Zappalà, L.; Mazzeo, G.; Siscaro, G. Lethal and sub-lethal effects of insecticides on natural enemies of citrus scale pests. *BioControl* **2009**, *54*, 651–661. [CrossRef]
52. Somasundaran, P.; Mehta, S.C.; Purohit, P. Silicone emulsions. *Adv. Colloid Interface Sci.* **2006**, *128-130*, 103–109. [CrossRef]
53. Hrastar, R.; Petrisic, M.G.; Ogrinc, N.; Kosir, I.J. Fatty acid and stable carbon isotope characterization of Camelina sativa oil: Implications for authentication. *J. Agric. Food Chem.* **2009**, *57*, 579–585. [CrossRef]
54. Copping, L.G.; Duke, S.O. Natural products that had been used commercially as crop protection agents. *Pest Manag. Sci,* **2007**, *63*, 524–554. [CrossRef]

55. Bakkali, F.; Averbeck, S.; Averbeck, D.; Idaomar, M. Biological effects of essential oils—A review. *Food Chem. Toxicol.* **2008**, *46*, 446–475. [CrossRef]
56. Skalský, M.; Niedobová, J.; Popelka, J. The efficacy of European fruit lecanium, *Parthenolecanium corni* (Bouché, 1844) control using natural products. *Hortic. Sci.* **2019**, *46*, 195–200. [CrossRef]
57. Gantner, M.; Jaśkiewicz, B.; Golan, K. Occurrence of *Parthenolecanium corni* (Bouché) on 18 cultivars of hazelnut. *Folia Hortic.* **2004**, *16*, 95–100.
58. Vončina, A.; Novljan, M. Effect of selected insecticides on European fruit lecanium population (*Parthenolecanium corni* Buche) in a northern highbush blueberry (*Vaccinium corymbosum* L.) orchard. In Proceedings of the 14. Slovensko Posvetovanje o Varstvu Rastlin z Mednarodno Udelezbo, Maribor, Slovenija, 5–6 March 2019; pp. 247–251.
59. Taylor-Wells, J.; Gross, A.D.; Jiang, S.; Demares, F.; Clements, J.S.; Carlier, P.R.; Bloomquist, J.R. Toxicity, mode of action, and synergist potential of flonicamid against mosquitoes *Pestic. Biochem. Physiol.* **2018**, *151*, 3–9.
60. Morita, M.; Ueda, T.; Yoneda, T.; Koyanagi, T.; Haga, T. Flonicamid, a novel insecticide with a rapid inhibitory effect on aphid feeding. *Pest Manag. Sci.* **2007**, *63*, 969–973. [CrossRef] [PubMed]
61. Karmakar, P.; Shera, P.S. Lethal and sublethal effects of insecticides used in cotton crop on the mealybug endoparasitoid *Aenasius arizonensis*. *Int. J. Pest Manag.* **2020**, *66*, 13–22. [CrossRef]
62. Serrão, J.E.; Plata-Rueda, A.; Martínez, L.C.; Zanuncio, J.C. Side-effects of pesticides on non-target insects in agriculture: A mini-review. *Sci. Nat.* **2022**, *109*, 17. [CrossRef] [PubMed]
63. Orr, D.B. Scelionid wasps as biological control agents: A review. *Fla. Entomol.* **1988**, *71*, 506–528. Available online: https://www.jstor.org/stable/3495011 (accessed on 10 March 2023). [CrossRef]
64. Sial, A. Identification and Management of Scale Insects in Blueberries. 2022. Available online: https://site.extension.uga.edu/ipm/2022/08/05/identification-management-of-scale-insects-in-blueberries/ (accessed on 10 March 2023).
65. Bahde, B.W.; Poojari, S.; Alabi, O.J.; Naidu, R.A.; Walsh, D.B. *Pseudococcus maritimus* (Hemiptera: Pseudococcidae) and *Parthenolecanium corni* (Hemiptera: Coccidae) are capable of transmitting Grapevine leafroll-associated virus between *Vitis* × *labruscana* and *Vitis vinifera*. *Environ. Entomol.* **2013**, *42*, 1292–1298. [CrossRef]
66. Hommay, G.; Komar, V.; Lemaire, O.; Herrbach, E. Grapevine virus A transmission by larvae of *Parthenolecanium corni*. *Eur. J. Plant Pathol.* **2008**, *121*, 185–188. [CrossRef]
67. Szyndel, M.S.; Paduch-Cichal, E. Detection of Blueberry red ringspot virus in different stages of *Parthenolecanium corni* in Poland. *Ann. Wars. Univ. LifeSci.–SGGW Hortic. Landsc. Archit.* **2020**, *41*, 77–81. [CrossRef]

Disclaimer/Publisher's Note: The statements, opinions and data contained in all publications are solely those of the individual author(s) and contributor(s) and not of MDPI and/or the editor(s). MDPI and/or the editor(s) disclaim responsibility for any injury to people or property resulting from any ideas, methods, instructions or products referred to in the content.

Article

Cladosporium Species: The Predominant Species Present on Raspberries from the U.K. and Spain and Their Ability to Cause Skin and Stigmata Infections

Lauren Helen Farwell [1,2,*], Greg Deakin [1], Adrian Lee Harris [1], Georgina Fagg [1], Thomas Passey [1], Carol Verheecke-Vaessen [2], Naresh Magan [2] and Xiangming Xu [1,*]

1 NIAB—New Road, East Malling, West Malling ME19 6BJ, UK
2 Applied Mycology Group, Environment and AgriFood Theme, Cranfield University, College Road, Bedford MK43 0AL, UK
* Correspondence: lauren.farwell@niab.com or lauren.farwell@cranfield.ac.uk (L.H.F.); xiangming.xu@niab.com (X.X.)

Abstract: Raspberry (*Rosales: Rosaceae*) production in the U.K. has moved rapidly in the last 10 years to under polythene, combined with a reduced availability of broad-spectrum fungicides. Hence, the incidence of previously less prevalent diseases, such as *Cladosporium* (*Capnodiales: Cladosporiaceae*), has largely increased. This study aimed to identify the predominant *Cladosporium* species on raspberry and to understand the nature of its infection on raspberry fruit. Raspberries were collected from farms across the U.K. and Spain and incubated; fungal isolates were then isolated from typical *Cladosporium* lesions and identified to the species level based on the sequences of the trans elongation factor α and actin genes. *Cladosporium cladosporioides* (Fres) de Vries was confirmed as the predominant species responsible for infecting raspberry fruit close to harvest on fruit from the U.K. and Spain, being present on 41.5% of U.K. fruit and 84.6% of Spanish fruit. Raspberries were subsequently inoculated at different developmental stages with *C. cladosporioides* isolates to determine the susceptibility to *Cladosporium* skin lesions and stigmata infections in relation to the developmental stage. Only the ripening and ripe raspberries were susceptible to *Cladosporium*, resulting in skin lesions. *Cladosporium* can colonise the stigmata of raspberries earlier in fruit development and future research is required to determine if such stigmata infections could cause subsequent skin lesion infections. This study has provided the necessary epidemiological information to develop effective management measures against the *Cladosporium* species.

Keywords: *Rubus ideaus*; epidemiology; inoculation; stigma; phylogenetics; susceptibility; ordinal regression

1. Introduction

Raspberry (*Rubus ideaus*; (*Rosales: Rosaceae*)) is an economically important soft fruit crop in the U.K. worth approximately £133 M/$171 M in 2020 [1], mostly grown under protection. Raspberries are delicate fruits susceptible to various fungal diseases, especially grey mould caused by *Botrytis cinerea* (*Helotiales: Sclerotiniaceae*). Based on recent epidemiological knowledge of raspberry grey mould on fruits under protection [2], a management strategy based on post-harvest cool-chain management without fungicides was developed and able to control grey mould effectively [3].

Recently, some U.K. raspberry growers reported significant losses in yield due to fruit rots caused by the *Cladosporium* (*Capnodiales: Cladosporiaceae*) species. Currently, there is insufficient epidemiological knowledge on raspberry fruit rot caused by the *Cladosporium* species to develop management strategies. One recent study in the U.S.A. examined the relationship between the *Cladosporium* species and *Drosophila suzukii* (*Diptera, Drosophilidae*) on raspberries; *C. cladosporioides*, *C. anthropophilum* Sandoval-Denis, Gené and Wiederhold and *C. pseudocladosporioides* Bensch, Crous and Braun were found to be the predominant species [4]. In the U.K., it is unknown which predominant *Cladosporium* species are causing raspberry fruit rot. Previous studies of *Cladosporium* raspberry rot have focused on post-harvest disease development [5]. For example, *Cladosporium* was found to be prevalent at one farm in Kent, U.K., where 55% of raspberries exhibited *Cladosporium* symptoms post-harvest [3]; however, the origin of this high post-harvest incidence of *Cladosporium* was unclear. In the U.S.A., Swett et al. [4] showed that *Cladosporium* was present on fruit pre-harvest and on under-ripe raspberries when harvested and then incubated for 4 days.

If *Cladosporium* can infect raspberry fruits pre-harvest, the level of infection may depend critically on fruit age, as was found for *Botrytis* [6] and powdery mildew [7] on strawberry. *Cladosporium cladosporioides* was reported to cause blossom blight of strawberries in Korea [8] and the U.S.A. [9]. There is no knowledge of whether the *Cladosporium* species can similarly infect raspberry flower tissues, such as the stigmata, and become established on the fruit surface pre-harvest, resulting in yield losses. There is, therefore, an urgent need to understand the predominant *Cladosporium* species on raspberries and the susceptibility of raspberry fruit at different developmental stages to infection by *Cladosporium*. This knowledge will facilitate better development of control strategies to minimise the impact of this fungal infection at both the pre- and post-harvest stage.

Thus, the objectives of this study were to (a) determine which were the predominant *Cladosporium* species present on U.K. and Spanish raspberry fruit; (b) investigate fruit susceptibility to *Cladosporium* in relation to fruit developmental stages; and (c) identify whether stigmata are susceptible to *Cladosporium* infection.

2. Materials and Methods

2.1. Determining Cladosporium Species Present on Raspberry Fruit

2.1.1. Isolate Collection

Raspberries were sent directly by U.K. growers or purchased from supermarkets in punnets (including fruit originating from Spain). The raspberries were incubated in polythene-bagged trays to maintain high humidity in ambient room conditions for four days to accelerate fungal growth. Fruit with typical *Cladosporium* lesions were dissected and the diseased parts were directly plated onto potato dextrose agar (PDA; Oxoid, Basingstoke, U.K., CM0139) in 9 cm Petri dishes (supplemented with 9.3 g per 100 mL NaCl to inhibit bacterial growth). The fungal outgrowth that resembled the *Cladosporium* species was then sub-cultured onto fresh PDA medium to obtain pure cultures.

2.1.2. DNA Extraction, Amplification and Sequencing

DNA from those candidate *Cladosporium* isolates were extracted using a rapid fungal DNA extraction method with Sigma extraction and dilution buffers [10]. An approximately 1 mm × 1 mm piece of mycelium was taken from each isolate and placed in 30 µL of extraction buffer (Sigma-Aldrich, Gillingham, UK, E7526) and heated to 95 °C for 5 min. Then, 30 µL of dilution buffer (Sigma-Aldrich, Gillingham, UK, D5688) was added, followed by 50 µL of sterile water. The final product was stored at -20 °C for subsequent downstream processes. Once the DNA was extracted, PCR was performed on the trans elongation factor α (TEF1α) (primer pair: EF1-728F and EF-2R; Table S1) and actin (ACT) (primer pair: ACT-512F and ACT-783R; Table S1) regions, as these have been shown to be accurate in identifying *Cladosporium* at the species level [11]. The PCR cycle for TEF1α was set as follows: 95 °C for 8 min, 40 cycles of 95 °C for 15 s, 55 °C for 20 s, 72 °C for 1 min and, finally, 72 °C for 5 min. For the ACT gene, it was set as follows: 94 °C for 5 min, 45 cycles of 94 °C for 45 s, 52 °C for 30 s, 72 °C for 90 s and, finally, 72 °C for 6 min, as described by Swett et al. [4]. The PCR products were run on a 1.5% agarose gel for 45 min at 100 V with a 1 kb base pair ladder. If strong bands were seen, the products were diluted 1:10 and then sent to Eurofins (Ebersberg, Germany) for Sanger sequencing. The chromatograms were trimmed to remove poor end regions, and any sequences with signs of contamination were removed. For each isolate, the forward and reverse reads were combined for each gene to create a consensus sequence using Geneious version 2019.2.1 (Auckland, New Zealand).

Individual consensus sequences were then run through the nrBLAST database using the megaBLAST version 2.12.0+ (Bethesda, MD, USA) programme for species identification based on the e-value (<10–80) and the percentage identity (>95%). If multiple species were found with similar probability, the species identity was resolved as ambiguous. In total, 54 isolates were sequenced for both ACT and TEF1α regions.

2.1.3. Phylogenetic Analysis

Sequences from the BLAST searches with high similarity (e-value < 10–80, percentage identity > 95%) were added to our query sequences in the phylogenetic analyses as reference sequences (Table S2). As some of the Centraalbureau voor Schimmelcultures (Westerdijk Institute, Utrecht, The Netherlands) *C. fusiforme* isolates were missing the 3′ locus of the TEF1α gene, they were excluded in the TEF1α tree. In total, 73 sequences were used for the ACT phylogenetic tree at 215 base pairs long and 71 isolates for the TEF1α tree at 508 base pairs long, with *Cercospora beticola* (CBS 116456) as the outgroup in both trees. Sequences were aligned separately for the ACT and TEF1α genes using the online version of MAFFT version 7 (Osaka, Japan) (https://mafft.cbrc.jp/alignment/server/index.html (accessed on: 27 October 2022); [12]) with the E-INS-I model. Alignments were visually inspected, and sequences were trimmed to remove the primers using the software MEGA version X (Philadelphia, PA, USA) [13]. MEGA was also used to build maximum likelihood trees with 1000 bootstrap replicates for the two gene regions, in which a Kimura 2-parameter evolutionary model [14] was used for the ACT and TEF1α regions.

2.2. Susceptibility of Fruit to Cladosporium Skin Lesions

2.2.1. Experimental Design and Treatments

Inoculation experiments were conducted in an experimental raspberry plantation in a polytunnel in late summer 2019 and 2020 at NIAB, East Malling in Kent, U.K., (Longitude: 51.288350, Latitude: 0.454766). Potted plants of one proprietary raspberry variety (one of the most popular commercial varieties grown in the UK on which *Cladosporium* was previously reported by U.K. growers) were used and grown with an industry-standard fertigation regime for this specific cultivar. All three *C. cladosporioides* isolates used for inoculations were from U.K. farms and included in the phylogenetic analyses.

Fruits were classified into three developmental stages (green, ripening and ripe berries (Figure 1)) at the time of inoculation. Inoculations were conducted three times each year. On each plant, all fruits from one or more randomly selected branches with the widest range of fruit developmental stages were inoculated on each inoculation occasion unless otherwise specified. Any visible rotting fruits were removed prior to inoculation.

Figure 1. The three stages (green, ripening and ripe fruit) of raspberry fruit development inoculated in 2020 to investigate skin lesion development and stigmata infections by *Cladosporium*.

In 2019, for each fruit developmental stage, there was only one inoculation treatment without an un-inoculated control; fruits were inoculated with a mixture of three *C. cladosporioides* isolates. Fruits were inoculated on 29 August 2019, 5 September 19 and 12 September 2019. There were 12 pots (each with two canes) that were placed into four rows of three pots. Each row of plants was considered as one block since the three pots were fertigated by the same fertigation line. All 12 plants were inoculated on all three occasions—only a single branch was inoculated in each pot.

In 2020, there were four inoculation treatments: three isolates (as used in 2019) and a control. Inoculations were conducted on 9 September 2020, 16 September 2020 and 23 September 2020. There were 32 pots, each with three canes, in two rows. Each row of 16 pots was divided into four blocks, with one block containing four consecutive pots. One of the four pots in each block was randomly assigned to one of the four inoculation treatments. In each pot, 2–3 branches with the widest range of developmental stages were randomly selected for inoculation on each inoculation date.

2.2.2. Fruit Inoculation

Cladosporium cladosporioides isolates were grown on PDA (Oxoid, Basingstoke, UK, CM0139) for 7–10 days. Thirty millilitres of 0.5% KCl sterile water solution with 0.1% Tween-20 was placed onto the colony surface and agitated with a surface-sterilised glass rod, as recommended by Swett et al. [4]. The conidial suspension was decanted into a 25 mL universal bottle, and the concentration was determined using a haemocytometer and then adjusted to 10^7 conidia/mL using sterile water. To make the mixed inoculum in 2019, the same volumes of the adjusted conidial suspensions for the three isolates were mixed together. For the control treatment in 2020, 0.5% KCl and 0.1% Tween-20 solution was used to inoculate fruits. Inoculations were carried out with a handheld sprayer to thoroughly wet individual fruit. Immediately after inoculation, each inoculated branch was covered with a polythene bag and plugged with a damp cotton bung. This was held in place with a cable tie to maintain high humidity inside the bag, which was left in place for 24 h and then removed. All the inoculated fruit on the branch were then removed and placed in separate trays.

In 2019, immediately after removal, fruits were surface-sterilised by gently washing the fruit in 0.1% Tween-20 for 2 min, 70% Ethanol for 30 s and, finally, 0.1% bleach for 2 min, then left in a laminar flow hood for approximately 30 min to dry, as described by Swett et al. [4], and placed in surface-sterilised trays for incubation at 20 °C in 24 h of light for four days before assessment.

In 2020, all the inoculated fruit on a single branch were removed into one punnet and then sterilised, as in 2019, but with an additional 2-min sterile distilled water wash at the end. Fruits were then placed into punnets, covered with a polythene bag to maintain high humidity and incubated at 20 °C in 24 h of light for four days.

2.2.3. Infection Assessment

In 2019, each individual fruit was assessed for its developmental stage and for the presence or absence of visual *Cladosporium* rot after incubation for four days. In 2020, each individual fruit was assessed for its developmental stage and the number of drupelets with visual *Cladosporium* symptoms on the following scale: 0—no drupelets infected, 1—one drupelet infected, 2—two drupelets infected, 3—three drupelets infected, 4—four drupelets infected, and 5—$\geq$ 5 drupelets infected.

2.2.4. Statistical Analyses

To estimate the effect of inoculation date, fruit development stage and their interaction on the number (incidence) of *Cladosporium* infections for the 2019 inoculation, a Binomial generalised linear mixed model (GLMM) was fitted to the data using the glmmTMB package [15] in R version 4.0.3 (Vienna, Austria) [16]. In GLMM, residual errors were assumed to follow a binomial distribution, assuming no over-dispersion. Row (block), date of inoculation and fruit development stage were added as fixed effect factors to the model. The plant (pot) was added as a random-effect factor to control for plant effects. An ordinal logistic regression model was fit to the 2020 skin lesion infection score data using the R "ordinal" package [17]. Five fixed-effect factors were used: block, inoculation date, *Cladosporium* isolate, *Cladosporium*-inoculated or not and fruit development stage. Fixed effects showed proportional odds, meeting the assumptions of ordinal regression. Green fruit (which all had no infection) were excluded from the analysis due to the complete separation causing converging issues with the model fit. The fitted ordinal model was used to generate estimated incidences of infections (either uninfected [score = 0] or infected [score $\geq$ 1]). The standard errors were also generated via the fitted ordinal model and adjusted by multiplying by the root of the estimated over-dispersion parameter. In all analyses, statistical significance was based on deviance analysis via a Chi-square test.

2.3. Susceptibility of Stigmata to Cladosporium Infection

2.3.1. Experimental Design

This experiment focused on stigmata infection in relation to fruit development stage (Figure 1) and surface-sterilisation, which was used to indicate if *Cladosporium* hyphae could penetrate stigmata or only colonise the stigmata surface. The same experimental planting was used as for the 2020 fruit inoculation experiment. Three inoculations were performed on: 9 September 2020, 16 September 2020 and 23 September 2020 (the number of raspberries assessed for each treatment are shown in Table S3). On each inoculation date, in addition to the fruit developmental stage and the four inoculation treatments (as for the skin lesion experiment), there was an additional treatment factor: with or without post-collection surface-sterilisation before incubation. Individual branches on each pot were randomly assigned to the "sterilised" or "unsterilised" treatment.

2.3.2. Fruit Inoculation

The same inoculation methods were used as described for the 2020 skin lesion experiments. All fruit on those selected branches were inoculated and picked 24 h after inoculation. When appropriate, fruit were surface-sterilised as described for the 2020 skin lesion susceptibility experiment; fruit from one branch were then placed into a single punnet covered with a polythene bag and incubated for a further four days at 20 °C before assessment.

2.3.3. Assessment

Individual fruit were scored for the severity of stigmata infections on the following scale: 0—no stigmata infected, 1—1–20% of stigmata infected, 2—21–40% of stigmata infected, 3—41–60% of stigmata infected, 4—61–80% of stigmata infected, and 5—81–100% of stigmata infected.

2.3.4. Statistical Analyses

An ordinal logistic regression was used to analyse the data as for the 2020 skin lesion data with one additional fixed-effect factor (sterilization vs. unsterilised). As all the green fruit scored 0, they were excluded from the analysis.

3. Results

3.1. Determination of Predominant Cladosporium Species on Raspberry Fruit

In total, molecular analyses of 54 isolates from raspberry fruit were performed, with 41 from the U.K. and 13 from Spain. Molecular analyses showed that in the U.K., the most frequently present *Cladosporium* species was *C. cladosporioides*, present on 41.5% of ripe fruit. The next most common species were *C. sphaerospermum* and *C. europaeum*, present on 14.6% of ripe fruit. On the U.K. fruit, 9.8% of isolates were identified as *C. limoniforme* and 7.3% as *C. ramotellum*. Finally, 12.2% of the isolates from U.K. grown raspberries gave a mixture of species from the BLAST database; however, their identities closely matched *C. fusiforme*.

Similarly, the predominant species present on fruit from Spain was *C. cladosporioides*, present on 84.6% of fruit. The only other species present was *C. sphaerospermum*, present on 15.4% of fruit (Table 1).

Table 1. The percentage breakdown of *Cladosporium* species present on raspberries collected from the U.K. and Spain. The total number of isolates was 54, with 41 from the U.K. and 13 from Spain.

Cladosporium Species	U.K. Fruit	Spanish Fruit
C. cladosporioides	41.5	84.6
C. fusiforme	12.2	0.0
C. sphaerospermum	14.6	15.4
C. europaeum	14.6	0.0
C. limoniforme	9.8	0.0
C. ramotellum	7.3	0.0

Phylogenetic analysis of both the ACT and TEF1α region corroborated with the species ID based on the nrBLAST database. Isolates showing close matches with *C. fusiforme* also closely matched a mixture of other *Cladosporium* species (however, with poor query coverage), which were also added to the phylogenetic analysis. These isolates, however, still formed a clade with the *C. fusiforme* reference sequences in the phylogenetic trees.

In both trees (Figures 2 and S1), a mixture of *Cladosporium* isolates were found in individual farms; but for those isolates closely related to *C. fusiforme*, three of the five isolates originated from one farm in Kent.

3.2. Fruit Susceptibility to Cladosporium Skin Lesions

In 2019, *Cladosporium* skin lesions were only present on fruit at the ripening and ripe stage. The overall incidence of fruit with *Cladosporium* skin lesions were 13.2% (s.e. ± 8.3%) on ripening fruit and 26.5% (s.e. ± 11.5%) on ripe fruit. There was no significant difference in the presence of *Cladosporium* on ripening and ripe fruit (X^2 (1, $n = 36$) = 1.06, $p = 0.303$).

In 2020, *Cladosporium* skin lesions were present only on fruit at the ripening and ripe stages, with the overall incidence being 50.2% (s.e. ± 12.1%) of ripe fruit with skin lesions and 17.4% (s.e. ± 9.0%) of ripening fruit with skin lesions. Ripe fruit were more susceptible to *Cladosporium* than ripening fruit: the odds of fruit with *Cladosporium* skin lesions at a given severity score (odds = infected/healthy) increased to 5.14 (s.e. ± 1.3) ($p < 0.001$) times when fruit moved from the ripening to ripe stage. Inoculation led to more fruit with skin lesions with an overall incidence of *Cladosporium* of 36% (s.e. ± 9.9%) in inoculated and 20.4% (s.e. ± 13.2%) in the control ($p < 0.05$). There was also a significant difference in *Cladosporium* skin lesion scores amongst the three isolates ($p < 0.05$): isolate three caused significantly higher skin lesion severity scores when compared to isolate two.

3.3. Infection of Stigmata by Cladosporium

Stigmata infections occurred across all fruit developmental stages irrespective of surface-sterilisation (Figure 3). There were significant differences in the severity scores between inoculation dates ($p < 0.001$), inoculated or uninoculated ($p < 0.001$), three *Cladosporium* isolates ($p < 0.001$) and surface-sterilisation ($p < 0.001$) on the severity of stigmata infections. There were no significant differences in stigmata infection scores across developmental stages ($p = 0.166$).

There were higher incidences of *Cladosporium* on stigmata across all the developmental stages in the unsterilised treatments compared to the sterilised treatments (Table 2).

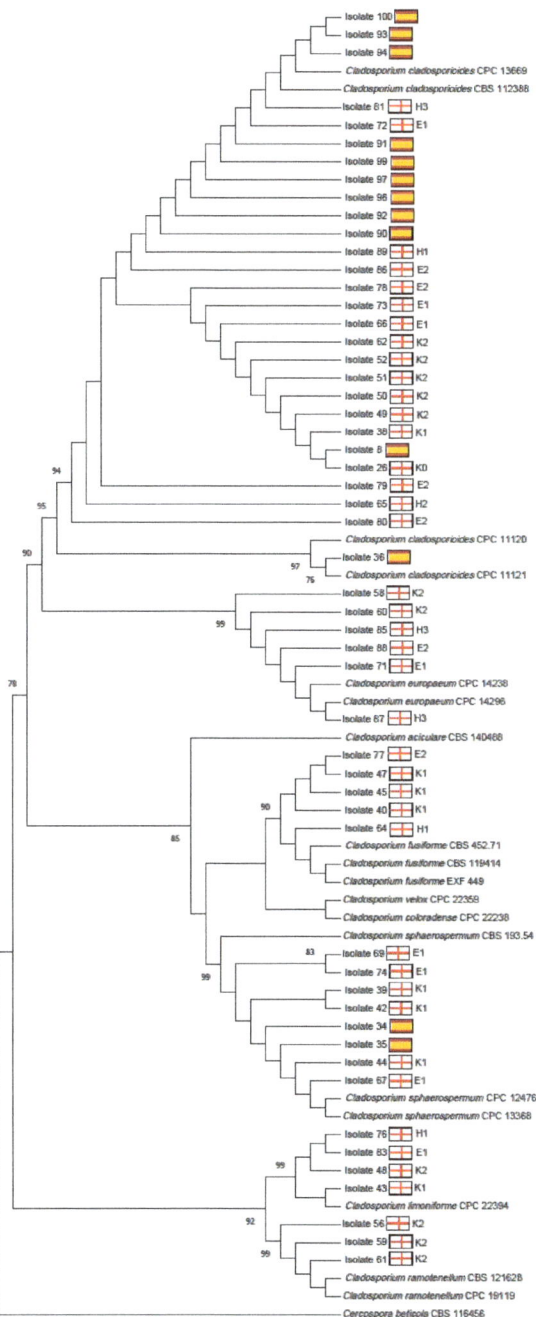

Figure 2. A phylogenetic tree of the ACT gene for 72 isolates of *Cladosporium*. The tree was created using maximum likelihood with a Kimura 2-parameter evolutionary model and 1000 bootstrap replications. The numbers above branches represent bootstrap values. The flags represent the country of origin, the codes represent areas within the country and numbers denote different farms. Area codes are as follows: E = Essex, K = Kent, H = Herefordshire.

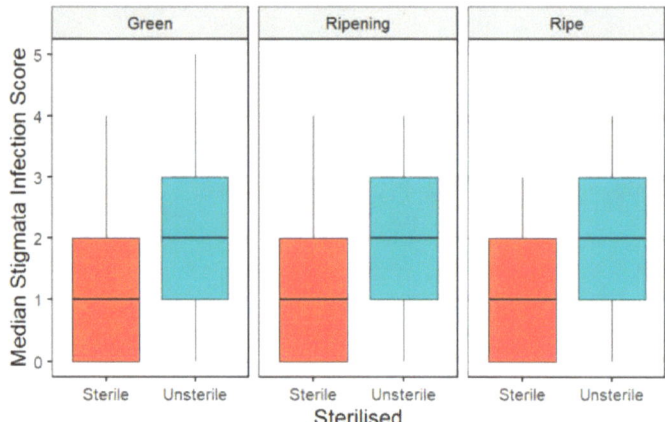

Figure 3. A boxplot of stigmata infection scores across the three developmental stages and between sterile vs. unsterile raspberry fruits; the black horizontal line is the median score.

Table 2. The overall incidence (% fruit with *Cladosporium* present) of *Cladosporium* on stigmata across four raspberry developmental stages and between sterilisation treatments. s.e.: standard error of the means.

Sterilisation Treatment	Green	Ripening	Ripe
Sterilised	66.4 (s.e. ± 6.9)	74.5 (s.e. ± 12.3)	69.9 (s.e. ± 16.5)
Unsterilised	88.3 (s.e. ± 6.8)	87.0 (s.e. ± 11.1)	90.6 (s.e. ± 8.6)

The incidences of *Cladosporium* infections on stigmata were also higher across all developmental stages inoculated with isolate three vs. isolate one and two (Table 3).

Table 3. The overall incidence (% fruit with *Cladosporium* present) of *Cladosporium* on stigmata across raspberry developmental stages and between inoculums. s.e.: standard error of the means.

Inoculum	Green	Ripening	Ripe
Control	65.7 (s.e. ± 10.5)	68.7 (s.e. ± 14.3)	70.1 (s.e. ± 14.7)
Isolate 1	79.1 (s.e. ± 9.1)	81.2 (s.e. ± 10.7)	82.2 (s.e. ± 10.8)
Isolate 2	81.0 (s.e. ± 8.7)	82.9 (s.e. ± 10.1)	83.9 (s.e. ± 10.7)
Isolate 3	87.9 (s.e. ± 6.4)	89.3 (s.e. ± 7.3)	89.9 (s.e. ± 7.2)

4. Discussion

This study has shown that *C. cladosporioides* is the predominant species infecting raspberry fruit in the U.K. and Spain, consistent with findings in the U.S.A. [4]. Several other *Cladosporium* species were also found on ripe raspberries, including *C. sphaerospermum*, *C. europaeum*, *C. limoniforme*, *C. ramotellum* and *C. fusiforme*. Of particular interest is the first isolation of *C. fusiforme* from plant material, suggesting the possibility of its presence on other crops. Three of the five isolates of this species were detected at one farm in Kent, indicating that this farm may provide an ecological niche favourable to this species. Some of these *Cladosporium* species were tested for their pathogenicity on other important horticultural crops, such as *C. limoniforme* on strawberries [18] and *C. cladosporioides*, *C. limoniforme* and *C. ramotellum* on pears and apples [19].

As *C. cladosporioides* was the most prevalent species found on raspberries, a large proportion of the primary inoculum may arise from airborne spores, as well as from crop debris [20], as this species comprises a large proportion of the airborne fungal load [21]. Previous studies investigating the patterns of *Cladosporium* airborne spores have focused on indoor and outdoor environments [22,23]. It may, therefore, be necessary to investigate how agricultural environments, e.g., inside polytunnels, impact the inoculum load and, hence, disease risk. There also appear to be differences among the three *C. cladosporioides* isolates in causing skin lesions and stigmata infections. Thus, in addition to inter-species differences in pathogenicity, further work is also needed to assess the intra-species variability in pathogenicity on raspberry fruit.

Present results indicate that ripe raspberries are most susceptible to *Cladosporium* skin lesions, followed by ripening fruit. In contrast, green fruit are not susceptible to *Cladosporium* skin lesions. Similarly, Swett et al. [4] showed that ripe raspberries are most susceptible to *Cladosporium* infection. Due to ripe fruit being delicate and susceptible to attacks by pests such as *Drosophila suzukii*, wounds may provide entry points for *Cladosporium* species, as found in the U.S.A. [4]. In addition, manual fruit harvesting may cause small abrasions to the surface, allowing access for fungi such as *Cladosporium* to infect and cause visible symptoms post-harvest.

Fruit from the green developmental stage onwards appear susceptible to *Cladosporium* stigmata infections. The severity of stigmata infections decreased when the fruit were surface-sterilised, indicating a significant proportion of *C. cladosporioides* inoculum only colonised the stigmata surface. Thus, the stigmata of raspberries are susceptible to *C. cladosporioides* and this species grows more readily as a saprophyte on the stigmata surface. The present results are consistent with a previous finding that *Cladosporium* rarely invaded the internal tissues of the grape fruiting structure [24]. *Cladosporium cladosporioides* can colonise or infect raspberry stigmata early in fruit development. However, further research is needed to study the dynamics of *Cladosporium* spore survival on the fruit surface. It is unknown if *Cladosporium* spores could survive long enough to cause secondary infections from the stigmata onto the skin's surface. Similar secondary infections were observed with *Botrytis cinerea* on raspberries [25]. *Cladosporium cladosporioides* was demonstrated to infect strawberry flowers [8]; therefore, similar investigations into the susceptibility of raspberry flowers to *Cladosporium* infections are needed.

Overall, it may be more economical to focus on managing *Cladosporium* infection when fruit are most susceptible to skin lesions, especially at the ripening stages of development. Improved husbandry and hygiene by removing decaying material may aid in reducing the inoculum load to prevent both skin and stigmata infections. The use of chemical fungicides to manage *Cladosporium* is limited unless there are effective products with a short half-life. As *Cladosporium* mainly grows on the stigmata surface, biological control agents may be able to compete effectively with *Cladosporium*, leading to much reduced incidences of raspberries with *Cladosporium* infections. The use of biocontrol agents against the *Cladosporium* species was demonstrated in multiple studies. Examples include multiple species of *Trichoderma* inhibiting the growth of *C. herbarum* [26] and *Chaetomium globosum* inhibiting the growth of multiple isolates of *C. cladosporioides* [27] during in vitro plate assays. Further research is needed to demonstrate the efficacy of such biological control agents in field conditions against the *Cladosporium* species.

Overall, the data obtained in this study will be beneficial in understanding the predominant *Cladosporium* species impacting raspberries and the growth stages of the ripening fruit most susceptible to *Cladosporium* infections. This will benefit agronomists and growers in more accurately assessing *Cladosporium* rot symptoms and identifying the most appropriate, and timing of, the application of control measures.

Supplementary Materials: The following supporting information can be downloaded at: https://www.mdpi.com/article/10.3390/horticulturae9020128/s1, Figure S1: A phylogenetic tree of the TEF1α gene for 70 isolates of *Cladosporium*; Table S1: The sequences of the ACT and TEF1α primers; Table S2: The accession numbers of the reference sequences used in the phylogenetic analyses; Table S3: The number of raspberries inoculated and sterilised at each inoculation date.

Author Contributions: Conceptualization, L.H.F., X.X. and N.M.; methodology, L.H.F., X.X., N.M., A.L.H. and T.P.; formal analysis, L.H.F., X.X., C.V.-V. and G.D.; investigation, L.H.F., A.L.H. and G.F.; resources, L.H.F., T.P. and G.D.; software, L.H.F., X.X. and G.D.; validation, L.H.F., X.X. and G.D.; writing—original draft preparation, L.H.F.; writing—review and editing, X.X., N.M. and G.D.; visualization, L.H.F. and G.D.; supervision, X.X. and N.M.; project administration, L.H.F.; funding acquisition, X.X. and N.M. All authors have read and agreed to the published version of the manuscript.

Funding: This research is funded by the Biotechnology and Biological Sciences Research Council (BBSRC) Collaborative Training Partnerships (CTP) for Fruit Crop Research in partnership with NIAB and Cranfield University, BBSRC (BB/T509073/1) and an industry partner with Berry Gardens Ltd.

Institutional Review Board Statement: Not applicable.

Informed Consent Statement: Not applicable.

Data Availability Statement: The data generated and analysed in this study are available on request from the corresponding author.

Acknowledgments: We would like to thank Sarah Cohen for her initial support in the DNA amplification and sequencing. We would also like to thank Richard Harnden and Andrius Kumstys (Berry Gardens Ltd.) for their support in collecting raspberry samples from across the U.K. Finally, we would like to thank Harriet Duncalfe for her advice and guidance during the experiments.

Conflicts of Interest: The authors declare no conflict of interest.

References

1. Department for Environment Food & Rural Affairs. Horticulture Statistics. 2020. Available online: https://assets.publishing.service.gov.uk/government/uploads/system/uploads/attachment_data/file/1003935/hort-report-20jul21.pdf (accessed on 26 November 2021).
2. Xu, X.; Wedgwood, E.; Berrie, A.M.; Allen, J.; O'Neill, T.M. Management of raspberry and strawberry grey mould in open field and under protection. A review. *Agron. Sustain. Dev.* **2011**, *32*, 531–543. [CrossRef]
3. O'Neill, T.; Wedgwood, E.; Berrie, A.M.; Allen, J.; Xu, X. Managing grey mould on raspberry grown under protection without use of fungicides during flowering and fruiting. *Agron. Sustain. Dev.* **2012**, *32*, 673–682. [CrossRef]
4. Swett, C.L.; Hamby, K.A.; Hellman, E.M.; Carignan, C.; Bourret, T.B.; Koivunen, E.E. Characterizing members of the *Cladosporium cladosporioides* species complex as fruit rot pathogens of red raspberries in the mid-Atlantic and co-occurrence with *Drosophila Suzukii* (Spotted wing drosophila). *Phytoparasitica* **2019**, *47*, 415–428. [CrossRef]
5. Dennis, C. Effect of pre-harvest fungicides on the spoilage of soft fruit after harvest. *Ann. Appl. Biol.* **1975**, *81*, 227–234. [CrossRef]
6. Jersch, S.; Scherer, C.; Huth, G.; Schlösser, E. Proanthocyanidins as basis for quiescence of *Botrytis cinerea* in immature strawberry fruits/Proanthocyanidine als ursache der quieszenz von *Botrytis Cinerea* in unreifen erdbeerfrüchten in unreifen erdbeerfrüchten. *zeitschrift für pflanzenkrankheiten und pflanzenschutz. J. Plant Dis. Prot.* **1989**, *96*, 365–378.
7. Carisse, O.; Bouchard, J. Age-related susceptibility of strawberry leaves and berries to infection by *Podosphaera aphanis*. *Crop Prot.* **2010**, *29*, 969–978. [CrossRef]
8. Nam, M.H.; Park, M.S.; Kim, H.S.; Kim, T.I.; Kim, H.G. *Cladosporium cladosporioides* and *C. tenuissimum* cause blossom blight in strawberry in Korea. *Mycobiology* **2015**, *43*, 354–359. [CrossRef]
9. Gubler, W.D.; Feliciano, A.J.; Bordas, A.C.; Civerolo, E.C.; Melvin, J.A.; Welch, N.C. First report of blossom blight of strawberry caused by *Xanthomonas fragariae* and *Cladosporium cladosporioides* in California. *Plant Dis.* **1999**, *83*, 400. [CrossRef]
10. Extract-N-Amp™ Plant PCR Kits Protocol. Available online: https://www.sigmaaldrich.com/GB/en/technical-documents/protocol/genomics/pcr/extract-n-amp-plant-pcr-kits-protocol (accessed on 18 March 2022).
11. Bensch, K.; Braun, U.; Groenewald, J.Z.; Crous, P.W. The genus *Cladosporium*. *Stud. Mycol.* **2012**, *72*, 1–401.
12. Katoh, K.; Rozewicki, J.; Yamada, K.D. MAFFT online service: Multiple sequence alignment, interactive sequence choice and visualization. *Brief. Bioinform.* **2019**, *20*, 1160–1166. [CrossRef]
13. Kumar, S.; Stecher, G.; Li, M.; Knyaz, C.; Tamura, K. MEGA X: Molecular evolutionary genetics analysis across computing platforms. *Mol. Biol. Evol.* **2018**, *35*, 1547–1549. [CrossRef] [PubMed]
14. Kimura, M. A simple method for estimating evolutionary rates of base substitutions through comparative studies of nucleotide sequences. *J. Mol. Evol.* **1980**, *16*, 111–120. [CrossRef] [PubMed]
15. Brooks, M.E.; Kristensen, K.; van Benthem, K.J.; Magnusson, A.; Berg, C.W.; Nielsen, A.; Skaug, H.J.; Mächler, M.; Bolker, B.M. GlmmTMB balances speed and flexibility among packages for zero-inflated generalized linear mixed modeling. *R J.* **2017**, *9*, 378–400. [CrossRef]
16. R Core Team. R Foundation for Statistical Computing. R: A Language and Environment for Statistical Computing. Available online: https://www.R-project.org/ (accessed on 26 November 2021).

17. Christensen, R.H.B. Ordinal—Regression Models for Ordinal Data. 2019. Available online: http://www.cran.r-project.org/package=ordinal/ (accessed on 29 November 2019).
18. Ayoubi, N.; Soleimani, M.J.; Zare, R. *Cladosporium* species, a new challenge in strawberry production in Iran. *Phytopathol. Mediterr.* **2017**, *56*, 486–493.
19. Temperini, C.V.; Pardo, A.G.; Pose, G.N. Diversity of airborne *Cladosporium* species isolated from agricultural environments of northern Argentinean Patagonia: Molecular characterization and plant pathogenicity. *Aerobiologia* **2018**, *34*, 227–239. [CrossRef]
20. Ellis, M.B. *Dematiaceous Hyphomycetes*, 1st ed.; Commonwealth Mycological Institute: Kew, UK, 1971; pp. 308–319.
21. Harvey, R. Air-spora studies at Cardiff: I. *Cladosporium*. *Trans. Br. Mycol. Soc.* **1967**, *50*, 479–495. [CrossRef]
22. Pasanen, A.L.; Niininen, M.; Kalliokoski, P.; Nevalainen, A.; Jantunen, M.J. Airborne *Cladosporium* and other fungi in damp versus reference residences. *Atmos. Environ.* **1992**, *26*, 121–124. [CrossRef]
23. Anees-Hill, S.; Douglas, P.; Pashley, C.H.; Hansell, A.; Marczylo, E.L. A systematic review of outdoor airborne fungal spore seasonality across Europe and the implications for health. *Sci. Total Environ.* **2022**, *818*, 151716. [CrossRef]
24. Briceño, E.X.; Latorre, B.A. Characterization of *Cladosporium* rot in grapevines, a problem of growing importance in Chile. *Plant Dis.* **2008**, *92*, 1635–1642. [CrossRef]
25. Kozhar, O.; Peever, T.L. How does *Botrytis cinerea* infect red raspberry? *Phytopathology* **2018**, *108*, 1287–1298. [CrossRef]
26. Barbosa, M.A.G.; Rehn, K.G.; Menezes, M.; Mariano, R.L.R. Antagonism of *Trichoderma* species on *Cladosporium herbarum* and their enzimatic characterization. *Braz. J. Microiol.* **2001**, *32*, 98–104. [CrossRef]
27. El-Dawy, E.G.A.E.M.; Gherbawy, Y.A.; Hussein, M.A. Morphological, molecular characterization, plant pathogenicity and biocontrol of *Cladosporium* complex groups associated with faba beans. *Sci. Rep.* **2021**, *11*, 14183. [CrossRef] [PubMed]

Disclaimer/Publisher's Note: The statements, opinions and data contained in all publications are solely those of the individual author(s) and contributor(s) and not of MDPI and/or the editor(s). MDPI and/or the editor(s) disclaim responsibility for any injury to people or property resulting from any ideas, methods, instructions or products referred to in the content.

Article

Biological Control of Downy Mildew and Yield Enhancement of Cucumber Plants by *Trichoderma harzianum* and *Bacillus subtilis* (Ehrenberg) under Greenhouse Conditions

Héctor G. Núñez-Palenius [1], Blanca E. Orosco-Alcalá [1], Isidro Espitia-Vázquez [1], Víctor Olalde-Portugal [2], Mariana Hoflack-Culebro [3], Luis F. Ramírez-Santoyo [1], Graciela M. L. Ruiz-Aguilar [1], Nicacio Cruz-Huerta [4] and Juan I. Valiente-Banuet [3,*]

[1] División de Ciencias de la Vida, Campus Irapuato-Salamanca, Universidad de Guanajuato, Agronomía, DICIVA-CIS, Ex Hacienda el Copal km 9, Carretera Irapuato-Silao, Irapuato 36500, Guanajuato, Mexico

[2] Biotecnología y Bioquímica, Centro de Investigación y de Estudios Avanzados del IPN (Cinvestav), Libramiento Norte Carretera Irapuato León Kilómetro 9.6, Carr Panamericana Irapuato-León, Irapuato 36821, Guanajuato, Mexico

[3] Tecnologico de Monterrey, Escuela de Ingeniería y Ciencias, Epigmenio González 500, San Pablo, Santiago de Querétaro 76130, Querétaro, Mexico

[4] Colegio de Postgraduados, Edafología, Campus Montecillo, Km. 36.5, Carretera México-Texcoco, Montecillo, Texcoco 56230, Estado de Mexico, Mexico

* Correspondence: valiente@tec.mx; Tel.: +52-442-592-6268

Abstract: The downy mildew disease of cucurbits is considered the most economically damaging disease of Cucurbitaceae worldwide. The causal agent, *Pseudoperonospora cubensis* (Berkeley & Curtis), may cause complete crop losses of cucurbits. Few commercial cucurbit cultivars are resistant to this disease. Commercially, *P. cubensis* is controlled primarily with synthetic fungicides that inhibit or eliminate the pathogen. Several biological agents have also been identified that provide some level of control. In our study, foliar applications of three strains of *Trichoderma harzianum* and two native strains of *Bacillus subtilis* were evaluated for the control of the disease on cucumber plants grown under commercial greenhouse conditions. The study was conducted using a completely randomized design with six individual treatments during two production cycles: fall 2015 and spring 2016. The response variables included disease incidence and severity, plant height, total yield, fruit quality, and weight. *B. subtilis* provided the best control over the incidence and severity of the disease in both production cycles. Interestingly, while *T. harzianum* was less effective at controlling the disease, it enhanced plant growth and productivity, and produced a higher number of better-quality fruits per plot. This increased yield with higher quality fruits may result in higher profit for the growers.

Keywords: beneficial microorganisms; *Cucumis sativus* L.; integrated pest management; *Pseudoperonospora cubensis* (Berkeley & Curtis); resistance; virulence

1. Introduction

Cucumber (*Cucumis sativus* L.) is the third most important vegetable crop produced under protected agriculture conditions in Mexico. Currently, 10 percent of the total greenhouse area is used for cucumber production, after tomatoes (70%) and bell peppers (16%). Under greenhouse conditions, the yield of cucumber plants is affected by several biotic and abiotic factors [1,2].

The cucumber crops are affected by the downy mildew disease of cucurbits. This disease is the most economically damaging disease of Cucurbitaceae worldwide [3]. The causal agent is *Pseudoperonospora cubensis* (Berkeley & Curtis), an obligate oomycete [4]. This pathogen may cause complete crop losses in cucumber, melon, watermelon, and pumpkin [5,6]. Over the past three decades, *P. cubensis* has resurged around the world. New genotypes, races, pathologists, and mating types have been identified [4,7,8]. During the last decade, the pathogen has become more detrimental, and currently, it causes greater disease severity. Weather factors affect the infection and disease development of the downy mildew. Foliar necrosis appears more quickly under hot and dry weather. However, low temperature and high humidity conditions do not stop the infection process [8,9]. The exact influence of these factors on the daily infection of the pathogen has not been fully determined [10].

P. cubensis has plant specialization that affects a wide range of Cucurbitaceae hosts. Pathogen virulence can be classified into pathogenic types based on their compatibility with the differential set of cucurbit hosts. The genetic basis of the specialization of the hosts of *P. cubensis* is not yet known. Nonetheless, the diversity and high virulence complexity of *P. cubensis* within the pathogen population indicate that host resistance is not effective in controlling downy mildew of the cucurbits for the available commercial Cucurbitaceae cultivars [3,4,11].

Control of the downy mildew disease of cucurbits requires an integrated approach that involves a combination of synthetic and biological fungicides, along with the introduction of resistant cultivars [3]. Currently, few commercial cultivars are resistant to the downy mildew disease. Thus, synthetic fungicides that inhibit or eliminate the pathogen are the primary method of control [7,12]. The widespread use of fungicides has created problems that include water and soil pollution, toxicity to animals and humans, and the generation of resistance by *P. cubensis* [13]. Recently, several antagonistic beneficial microorganisms of the pathogenic fungus have been identified. Among them, several species and strains of the genus *Trichoderma* spp. and *Bacillus subtilis* have been shown to control downy mildew under experimental laboratory conditions [14]. *Trichoderma* spp. has been reported to increase plant immunity against invasive pathogens [15]. The microorganisms used for the biological control of downy mildew present different modes of action for pathogen contention. These mechanisms include mycoparasitism, competition for space and nutrients, induced systemic resistance (ISR), and antibiosis mediated by the secretion of cell wall degrading enzymes. Reports on *Trichoderma* and *Bacillus subtilis* indicate that both microorganisms use all these mechanisms to control fungal diseases in plants under in vitro and greenhouse conditions [16,17].

The objective of this research was to evaluate the effectiveness of three strains of *Trichoderma* spp. and two of *Bacillus subtilis* for the control of *Pseudoperonospora cubensis* (downy mildew of cucurbits) and their effects on the yield and quality of cucumber crops grown under commercial greenhouse production conditions.

2. Materials and Methods

2.1. Study Area

The experiments were carried out under protected agriculture conditions in a plastic greenhouse located in Jaral del Progreso, Guanajuato (20.37° N, 101.067° W, and altitude 1735 m). This area has a humid subtropical climate according to the Köppen–Geiger weather classification system. Average temperatures are 18.5 °C, with minimum and maximum temperatures of 5 °C and 35.2 °C, respectively. The annual average rainfall is 687 mm, with February as the driest month (7 mm on average) and August with the highest precipitation (148 mm on average).

2.2. Crop Management and Application of Microorganisms

Two different cucumber crops were established during the fall–winter 2015–2016 (FW) and the spring–summer 2016 (SS) production cycles. The cucumber cultivars used for this study were the American type 'Paraiso' for the FW cycle and the Persian type 'Kathrina' for the SS cycle (Enza Zaden, http://www.enzazaden.com.mx, accessed on 15 September 2015). Seeds were planted in 50 cavity trays in August and March for the FW and SS cycles, respectively. Plants were transplanted 15 d later directly into the soil of the greenhouse. The greenhouse soil was a clay loam texture with a pH of 7.36, electrical conductivity of 2.32 dS·m^{-1}, and 2.04% of organic matter. The soil contained 92.3 ppm of P, 25.7 ppm of NO_3^-, 597 ppm of K^+, 3244 ppm of Ca, 896 ppm of Mg, and 237 ppm of Na. Both cultivars were transplanted at a 2 m distance between rows and 0.4 m between plants in a double row, at a planting density of 2.5 plants·m^{-2}. Plant nutrition was administered using the Steiner nutrient solution using: $Ca(NO_3)_2·4H_2O$, KNO_3, $MgSO_4·7H_2O$, K_2SO_4, KH_2PO_4, and H_2SO_4. The concentrations of anions and cations of each nutrient expressed in $mol_c·m^{-3}$ present in the Steiner solution are shown in Table 1 [18]. Fertilizer solutions were applied daily at rates of 0.5 and 1.2 L·$plant^{-1}$ from 10 DAT to first anthesis, and from the first flower onward, respectively.

Table 1. Steiner nutrient solution.

Ions [1]	Cations molc m^{-3}	Anions molc m^{-3}	Total Ions
K^+	7		
Ca^{2+}	9		
Mg^{2+}	4		20
NO_3^-		12	
SO_4^{2-}		7	20
$H_2PO_4^-$		1	

[1] Ions needed to make the Steiner solution after the ions naturally occurring in the irrigation water were considered. The osmotic potential of the solution was 0.072 MPa, and the EC value was 2.0 dS m^{-1}. The commercial fertilizers used in the nutrient solution were $Ca(NO_3)_2·4H_2O$, KNO_3, $MgSO_4·7H_2O$, K_2SO_4, KH_2PO_4, and H_2SO_4.

All experimental plants were treated using the commercial practices for the control of the downy mildew disease used by the growers (biweekly applications of Serenade max®, Apolo®, and hydrogen peroxide (Q Basic®). The biological control treatments consisted of two native strains of *Bacillus subtilis* (VOB1 and VOB2) and three strains of *Trichoderma* spp. (VOT1, QLT, and BKNT). Both VOB1 and VOB2 *B. subtilis* strains and the VOT1 *Trichoderma* spp. strain were donated by Dr. Víctor Olalde (CINVESTAV, Unidad Irapuato). QLT was obtained from QLT by Química Lucava S.A. de C.V. (Grupo Lucava, http://grupolucava.com, accessed on 9 May 2015), and BKNT from Biokrone S.A. de C.V., (Biokrone, http://www.biokrone.com, accessed on 24 June 2015). Biological control treatments were also applied on a biweekly basis.

The biological treatments were applied by determining the dose of each strain for each treatment, and for each application, the strain or product was diluted in 5 L of water [19]. Also, Cosmocel®, a penetrating surfactant INEX-A®, was included (1 mL·L^{-1}) (Table 2). The solution was sprayed manually using a number three conical nozzle. All treatments were applied weekly during the phenological cycle of the crop.

Table 2. Treatments applied in cucumber cultivation in the Fall-Winter (2015–2016) and Spring-Summer (2016) cycles.

Treatment		Inoculum (Active Ingredient)	Dosage	Concentration
	(a) Serenade max®	*Bacillus subtilis*	0.8 g·L^{-1}	1×10^9 UFC·g^{-1}
		Bacillus subtilis		1×10^8 UFC·g^{-1}
Control †	(b) Apolo®	*Trichoderma harzianum*	0.8 g·L^{-1}	1×10^7 esp·g^{-1}
		Trichoderma viridae		1×10^7 esp·g^{-1}
		Streptomyces lydicus		1×10^8 UFC·g^{-1}
	(c) Hydrogen peroxide	H_2O_2	0.4 mL·L^{-1}	50%
VOT1		*Trichoderma harzianum*	0.8 g·L^{-1}	1×10^7 UFC·g^{-1}
QLT		*Trichoderma harzianum*	0.32 g·L^{-1}	1×10^7 UFC·g^{-1}
BKNT		*Trichoderma harzianum*	0.32 g·L^{-1}	1.1×10^7 UFC·g^{-1}
VOB1		*Bacillus subtilis*	0.8 mL·L^{-1}	1×10^9 UFC·mL^{-1}
VOB2		*Bacillus subtilis*	0.8 mL·L^{-1}	1×10^7 UFC·mL^{-1}

† Control, biweekly applications of commercial products as applied by the growers. Control treatments consisted of Serenade Max® (Bayer: strain QST 713 4.6%. Wettable powder (PH) 1×10^9 UFC·g^{-1}), Apolo® (Arvensis, https://arvensis.com.mx/, accessed on 19 August 2015, *Bacillus subtilis* 1×10^8 UFC·g^{-1}; *Trichoderma harzianum* 1×10^7 esp·g^{-1}; *Trichoderma viridae* 1×10^7 esp·g^{-1}; *Streptomyces lydicus* 1×10^6 UFC·g^{-1}; plant extracts 60.0% p·p^{-1}; Si 2.0% p·p^{-1}), and hydrogen peroxide (Q Basic). UFC: unit-forming colonies.

2.3. Experimental Design

The study was conducted in a commercial greenhouse in which *Pseudoperonospora cubensis* was prevalent. The experimental design consisted of a completely randomized design with six treatments (T) (Table 2) and 100 randomly distributed repetitions per treatment. The biological control strains for each treatment were assigned randomly in both production cycles (FW and SS). The harvest dates of the FW crop cycle for the 'Paraiso' cultivar were at 30, 60, 90, and 120 days after transplant (dat), while the 'Kathrina' cultivar during the SS cycle were at 30, 60, and 90 dat. The experimental units consisted of individual cucumber plants in each of the production systems. Due to the severity of the disease, a control treatment with no chemical applications was not viable for a study under commercial conditions. This situation caused the need for control treatments to reduce the expansion of the disease, senescence of the plant, and yield loss. These treatments allowed us to identify the effectiveness of the proposed microorganisms under commercial conditions in the greenhouse.

The downy mildew disease was evaluated by classifying cucumber plants from each treatment by the level of disease symptoms according to the method described by Ruiz Sánchez et al. (2008) [20]. Disease data were taken at 30, 60, 90, and 120 dat for the FW cycle, and at 30, 60, and 90 dat for the SS cycle, due to the duration of the production cycles. Disease incidence was determined by counting the number of plants with symptoms relative to the total number of plants in each experimental plot. A severity scale was developed using the Horsfall–Barratt method. This method is based on assigning a numerical value based on the percentage of foliar area with disease symptoms. In our study, these percentages were t: 1 = 0%, 2 = 0–3%, 3 = 3–6%, 4 = 6–12%, 5 = 12–25%, 6 = 25–50%, 7 = 50–75%, 8 = 75–88%, 9 = 88–94%, 10 = 94–97%, 11 = 97–100%, 12 = 100% [21] (Figure 1).

Figure 1. Horsfall–Barratt method for the establishment of severity index of downy mildew in cucumber leaves. Percentage represents the fraction of damage in the leaves.

The yield of cucumber plants for each treatment was determined by harvesting and weighing the cucumbers during the phenological cycle. Harvesting for the 'Kathrina' Persian-type finished at 90 dat and for the 'Paraiso' American-type cucumber at 120 dat. Fruit quality was determined using the standards of a commercial packinghouse (INTEBAJ, http://www.intebaj.com/, accessed on 17 May 2015) (Table 3). Plant height was measured from the base to the apex of the plant.

Table 3. Quality standards for cucumber fruits of the INTEBAJ commercial packinghouse.

Size [†]	'Paraiso'			'Kathrina'		
	First	Second	Third	First	Second	Third
Length (cm)	13.5–15	12–13.5, 15–17	<12, >17	23–25	18–23, 25–30	<18, >30
Width (cm)	3.3–3.5	3–3.3, 3.5–4	<3, >4	5.6–6	5–5.6, 6–6.6	<5, >6.6
Curvature (degree)	0°	10–20°	>20°	0°	20–30°	>30°
Damages (%)	0	<30	>30	0	<30	>30

[†] Based on cucumber fruit quality standards of Terra Bella (California, USA).

2.4. Statistical Analysis

The statistical analysis performed for the variables of yield per cucumber plant, plant height, and individual weight of the cucumbers in a factorial design with a completely randomized design was an analysis of variance, followed by the comparison of means by the Tukey method ($\alpha = 0.05$). Disease severity was evaluated using the Kruskal–Wallis test, followed by the Dunn's method for the comparison of means ($p \leq 0.05$). All analyses were carried out using the statistical analysis system program (SAS Institute, Cary, NC, USA).

3. Results and Discussion

The degree of disease severity was significantly different for both the American and the Persian type cucumbers at 60 dat. The treatments with the best controlling effect were *B. subtilis* VOB1, *B. subtilis* VOB2, and *T. harzianum* QLT, followed by the *T. harzianum* VOT1 strain at 60, 90, and 120 dat. These same strains also showed adequate disease control at 60 and 90 dat for the SS cycle (Table 4). By contrast, the plants that had the highest incidence of downy mildew in cucumber plants were the control and the BKNT strains treatments.

Table 4. Effect of treatments on the degree of severity of downy mildew on cucumber crops during the FW and SS cycles.

Treatment [†]	Cycle FW ('Paraiso')				Cycle SS ('Kathrina')		
	30 dat	60 dat	90 dat	120 dat	30 dat	60 dat	90 dat
Control	1.52 a	2.08 ab	2.2 b	2.49 b	1.58 a	4.57 a	7.91 a
VOT1	1.51 a	1.83 bc	2.0 bc	2.21 bc	1.54 a	3.69 bc	6.15 b
QLT	1.50 a	1.76 c	1.9 c	2.25 bc	1.55 a	3.52 c	6.34 b
BKNT	1.54 a	2.36 a	2.9 a	3.10 a	1.58 a	4.06 ab	
VOB1	1.50 a	1.71 c	1.8 c	2.10 c	1.51 a	3.42 c	6.05 b
VOB2	1.50 a	1.64 c	1.9 c	2.17 c	1.51 a	3.31 c	5.77 b

[†] Control, biweekly applications of commercial products as applied by the growers. Control treatments consisted of Serenade Max® 14.6; Apolo®; H_2O_2; VOT1: *Trichoderma harzianum* strain VOT1; QLT: *T. harzianum* strain QLT; BKNT: *T. harzianum* strain BKNT; VOB1: *Bacillus subtilis* strain VOB1; VOB2: *B. subtilis* strain VOB2. Severity scale: 1 = 0%, 2 = 0–3%, 3 = 3–6%, 4 = 6–12%, 5 = 12–25%, 6 = 25–50%, 7 = 50–75%, 8 = 75–88%, 9 = 88–94%, 10 = 94–97%, 11 = 97–100%, 12 = 100% [21]. The data within the columns with different letters show significant differences in the Dunn's test ($p \leq 0.05$). Dat = days after transplant.

The highest disease severity of downy mildew was observable in the cucumber plants of the control treatments of both American- and Persian-type cucumbers (FW and SS cycles, respectively). In the control treatments, no microorganism types were applied (neither strains of *Trichoderma* and *Bacillus* nor BKNT of *T. harzianum*). Increased severity of the disease could be observed in the 'Kathrina' cucumber plants, which indicates their low resistance to the presence of *P. cubensis* (Figure 2). These findings are consistent with previous reports in which plant pathogens can be controlled using microbial antagonists [16].

'Kathrina' cultivar plants show the greatest disease severity compared to the plants of the cultivar 'Paraiso' (Table 4). These different susceptibilities could be related to genomic differences between the cultivars. Environmental conditions may also have had an important effect on disease severity as summer was warmer and more humid than the fall, which was drier.

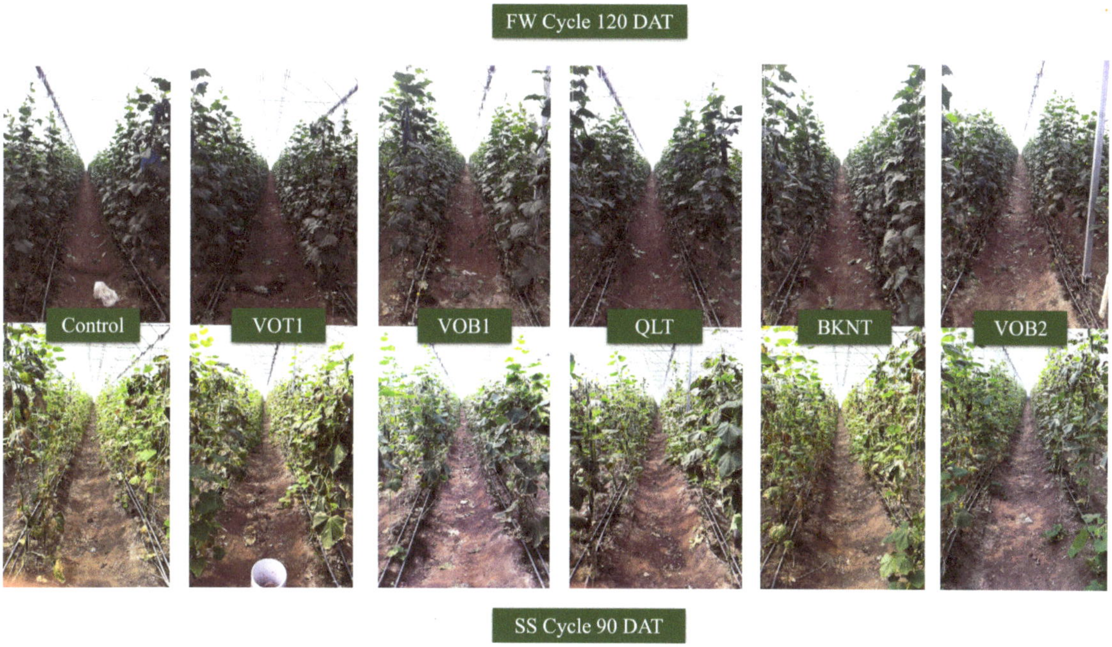

Figure 2. Greenhouse view of each treatment at the end of the fall–winter 2015–16 and spring–summer 2016 production cycles. The treatments were control: Serenade Max® 14.6; Apolo®; hydrogen peroxide); VOT1: *Trichoderma harzianum* strain VOT1; QLT: *T. harzianum* strain QLT; BKNT: *T. harzianum* strain BKNT; VOB1: *Bacillus subtilis* strain VOB1; VOB2: *B. subtilis* strain VOB2.

According to the severity scale developed for downy mildew of cucurbits [21], the VOB1 strain of *B. subtilis* presents the best control of the disease. Similar data were obtained for the VOB2 of *B. subtilis* for the FW and SS cycles.

Bacillus subtilis is considered a broad-spectrum disease-resistant microorganism capable of controlling different strains of pathogens of cucurbits [22]. The suppressive effects on plant pathogens by *B. subtilis* could be related to several mechanisms, including antibiosis, secretion of degrading enzymes, and competition for space and nutrients. *B. subtilis* might also induce the plants to generate systemic resistance and have other positive effects such as enhanced nutrient absorption (mainly N uptake), phosphate solubilization, production of phytohormones and siderophores, and increased plant growth. Enhanced plant nutrient absorption caused by *B. subtilis* may increase the capacity to tolerate the infection. Resistance may be improved by enzymes, or other metabolites independent of the direct action of *B. subtilis* on the pathogen. These factors might influence the improvement in the resistance of the cultivars to colonization by the pathogen [22,23]. To fully understand the mechanism by which *B. subtilis* enhances disease resistance, future studies should consider determining the expression of plant defense resistance genes.

By contrast, the treatment that presents the least amount of control over the downy mildew in the FW cycle is the BKNT strain *of T. harzianum*; and for the SS cycle, the BKNT strain is comparable to the control (Table 5). This indicates the susceptibility of the pathogen to strains of *B. subtilis*, but not to *T. harzianum*. Therefore, the genetic resistance of the host is not effective for the control of the mildew [4].

Fruit yield of cucumber plants is significantly different for the American and Persian types. Interestingly, while the *T. harzianum* VOT1 strain is not the strain that provides the best disease control, it causes a yield increase during both production cycles (FW and SS). (Table 6).

Table 5. Severity percentage of the downy mildew of cucurbits in cucumber crop, during the FW and SS cycles, according to the Horsfall–Barratt method [21] for the development of a severity scale.

Treatment [†]	Cycle FW ('Paraiso')				Cycle SS ('Kathrina')		
	30 dat	60 dat	90 dat	120 dat	30 dat	60 dat	90 dat
Control	1	29	45	62	5 *	88 *	100 *
VOT1	1	29	43	61	3	78	95
QLT	0	23	37	60	3	80	97
BKNT	2 *	43 *	85 *	100 *	5 *	85	100 *
VOB1	0	18	31	52	1	77	90
VOB2	0	13	31	54	1	76	90

[†] Control, biweekly applications of commercial products as applied by the growers. Control treatments consisted of Serenade Max® 14.6; Apolo®; hydrogen peroxide); VOT1: *Trichoderma harzianum* strain VOT1; QLT: *T. harzianum* strain QLT; BKNT: *T. harzianum* strain BKNT; VOB1: *Bacillus subtilis* strain VOB1; VOB2: *B. subtilis* strain VOB2. * = greater incidence percentage. Dat = days after transplant.

Table 6. Fruit yield of cucumber plants during FW and SS cycles.

Treatment [†]	Yield (kg·m^{-2})	
	Cycle FW ('Paraiso')	Cycle SS ('Kathrina')
Control	10.81 bc	8.11 c
VOT1	12.02 a	12.35 a
QLT	10.72 c	8.84 b
BKNT	10.89 bc	8.25 c
VOB1	11.05 b	8.84 b
VOB2	10.16 d	9.04 b

[†] Control, biweekly applications of commercial products as applied by the growers. Control treatments consisted of Serenade Max® 14.6; Apolo®; hydrogen peroxide); VOT1: *Trichoderma harzianum* strain VOT1; QLT: *T. harzianum* strain QLT; BKNT: *T. harzianum* strain BKNT; VOB1: *Bacillus subtilis* strain VOB1; VOB2: *B. subtilis* strain VOB2. Data in the columns with different letters indicate significant differences in the Tukey test ($p \leq 0.05$).

The VOT1 strain of *T. harzianum* generates the largest cucumber plants in both production cycles. In addition, VOT1-treated plants produce the largest individual fruit weights and total yield of cucumber plants, and the greatest number of fruits per harvest. In the SS cycle, the increase in yield of the VOB2 treatment is 36% higher than the control (Table 7).

Table 7. Effect of treatments on height (cm) of cucumber plants during the FW and SS cycles.

Treatment [†]	FW Cycle ('Paraiso')				SS Cycle ('Kathrina')		
	30 dat	60 dat	90 dat	120 dat	30 dat	60 dat	90 dat
Control	35.4 bc	88.6 b	134.8 b	187.9 b	129.1 bc	209.0 b	288.1 bc
VOT1	41.2 a	93.2 a	144.2 a	196.5 a	136.3 a	223.6 a	314.8 a
QLT	38.2 ab	88.8 b	132.6 b	186.0 b	140.0 abc	206.8 b	290.0 bc
BKNT	38.7 a	88.5 b	135.7 b	189.5 b	134.2 ab	211.5 b	291.6 b
VOB1	33.2 c	81.8 c	130.1 b	183.1 b	129.0 bc	206.1 b	290.3 b
VOB2	32.4 c	83.7 c	129.3 c	182.9 b	127.4 c	207.2 b	284.3 b

[†] Control, biweekly applications of commercial products as applied by the growers. Control treatments consisted of Serenade Max® 14.6; Apolo®; hydrogen peroxide); VOT1: *Trichoderma harzianum* strain VOT1; QLT: *T. harzianum* strain QLT; BKNT: *T. harzianum* strain BKNT; VOB1: *Bacillus subtilis* strain VOB1; VOB2: *B. subtilis* strain VOB2. Data in the columns with different letters indicate significant differences in the Tukey test ($p \leq 0.05$). Dat = days after transplant.

The VOB1 and VOB2 strains of *B. subtilis* cause the lowest growth of cucumber plants. The treatments do not significantly affect the number of fruits per harvest. Nevertheless, the VOT1 treatment produces the largest number of fruits per harvest (102.9 fruits at 120 dat in the FW cycle and 152.4 fruits at 90 dat in the SS cycle), followed by the VOB1 strain. The increase in number of fruits per cut induced by the VOT1 strain is 7.5% in the FW cycle and 33% in the SS cycle when compared to the control treatments.

The VOT1 treatment causes plants to produce fruits with the largest individual weights, even at the first harvests, which has a direct impact on cucumber plant yield (Table 8).

Table 8. Effect of treatments on weight (g) of individual cucumber fruits during the FW and SS cycles.

Treatment [†]	Cycle FW ('Paraiso')		Cycle SS ('Kathrina')		
	90 dat	120 dat	30 dat	60 dat	90 dat
Control	346.1 c	347.1 c	95 e	95 e	94.5 d
VOT1	359.6 a	359.6 a	108 a	108 a	108 a
QLT	350 b	349.4 b	98 d	98 d	97.9 c
BKNT	343 d	343.6 d	95 f	95 f	95 d
VOB1	346.5 c	347.8 bc	98 c	98 c	97.9 c
VOB2	340 d	340 e	99 b	99 b	99 b

[†] Control, biweekly applications of commercial products as applied by the growers. Control treatments consisted of Serenade Max® 14.6; Apolo®; hydrogen peroxide); VOT1: *Trichoderma harzianum* strain VOT1; QLT: *T. harzianum* strain QLT; BKNT: *T. harzianum* strain BKNT; VOB1: *Bacillus subtilis* strain VOB1; VOB2; *B. subtilis* strain VOB2. Data in the columns with different letters indicate significant differences in the Tukey test ($p \leq 0.05$). Dat = days after transplant.

However, the higher yield of cucumber plants of the VOT1 strain is more related to the larger number of fruits than to their individual fruit weights. In the case of 'Kathrina', the number of fruits increases by 33%, compared to an increase of 8% in their weights (Table 9). Previous studies reported similar differences in yield and quality due to changes in the use of varieties during different cycles, even within the same production system [24]. As for the fruit quality variable, we did not find a consistent response in both production cycles (SS and FW) because the quality classification for the type of cucumber (Persian or American) had a considerable influence on our results as quality standards are more rigorous for American than for Persian cucumber.

Table 9. Effect of the treatments on the first ('Premium') and second quality cucumber fruits during the FW and SS cycles.

Treatment [†]	Yield (kg m^{-2})									
	Cycle FW ('Paraiso')				Cycle SS ('Katrina')					
	90 dat		120 dat		30 dat		60 dat		90 dat	
	1st	2nd	1st	2nd	1st	2nd	1st	2nd	1st	2nd
Control	95.2 e	4.9 a	91.6 a	8.4 a	70.8 a	26 a	73.9 d	23.7 a	74.4 d	22.9 a
VOT1	99.1 a	0.1 e	95.9 a	4.1 c	78 a	20 a	83.9 a	14.3 b	80.3 a	13.9 c
QLT	97.6 bc	2.4 cd	94.8 a	5.2 ab	75.3 a	22.3 a	77.7 bcd	20.4 a	78.1 bc	20.0 a
BKNT	95.9 de	4.1 ab	92.8 a	7.2 ab	70.8 a	25.8 a	75.4 cd	22.9 a	75.9 cd	22.2 a
VOB1	96.6 cd	3.4 bc	93.4 a	6.8 b	75.5 a	21.8 a	78.5 bc	19.2 a	78.9 bc	18.8 b
VOB2	98.4 ab	1.5 de	95.2 a	4.8 bc	76.3 a	21.5 a	80 ab	18.5	80.3 b	18.3 b

[†] Control, biweekly applications of commercial products as applied by the growers. Control treatments consisted of Serenade Max® 14.6; Apolo®; hydrogen peroxide); VOT1: *Trichoderma harzianum* strain VOT1; QLT: *T. harzianum* strain QLT; BKNT: *T. harzianum* strain BKNT; VOB1: *Bacillus subtilis* strain VOB1; VOB2; *B. subtilis* cepa VOB2. Data in the columns with different letters indicate significant differences in the Tukey test ($p \leq 0.05$). Dat = days after transplant.

In addition, adverse weather conditions during the FW cycle caused greenhouse damage and affected the final stage of the crop. This condition caused a reduction in fruits of 'premium' quality and no statistically significant differences were found in the treatments at 120 dat. Nevertheless, the most notable strains were the VOT1 of *T. harzianum* and the VOB2 de *B. subtilis*, which caused a similar response, with the exception that the latter case was at 90 dat during the SS cycle (Table 9).

The commercial value (price) of first quality cucumber (or 'premium') can be up to 50–100% higher than those fruits of second quality. Therefore, the economic profit of cucumber cultivation is directly related to the quantity and quality of the obtained fruits.

In both cycles, the VOT1 strain of *T. harzianum* produced a higher quantity of 'premium' fruits in the SS cycle and a reduced number of second quality fruits. This higher quality crop represents a greater economic gain for the producer since, for the FW cycle, 95% of the 120 tons were of first quality, while for the SS cycle, 80% of 123 tons were also of prime quality. In the FW cycle, plants treated with synthetic fungicides had a yield of 108 t, of which 91% were of first quality, while during the SS cycle 81 t was obtained, 74% of which were of first quality. The treatments that presented the greatest amount of second quality fruits were the control, QLT, and BKNT. The latter (QLT and BKNT) were treated with strains of *T. harzianum* (Table 9). The effects of the treatments on third quality fruits were not significant (Table 10).

Table 10. Effect of treatments on third quality cucumber fruits during the FW and SS cycles.

Treatment *	Yield (kg m^{-2})				
	Cycle FW (Paraiso)		Cycle SS ('Kathrina')		
	90 dat	120 dat	30 dat	60 dat	90 dat
Control	1.69 a	2.787 a	1.813 a	1.514 a	1.338 a
VOT1	2.14 a	3.087 a	1.437 a	1.341 a	1.136 a
QLT	2.78 a	4.589 a	1.553 a	1.783 a	1.643 a
BKNT	2.59 a	3.9 a	1.609 a	1.507 a	1.444 a
VOB1	2.54 a	4.424 a	3.21 a	1.957 a	1.767 a
VOB2	2.67 a	4.91 a	1.525 a	1.454 a	1.371 a

* Control, biweekly applications of commercial products as applied by the growers. Control treatments consisted of Serenade Max® 14.6; Apolo®; hydrogen peroxide); VOT1: *Trichoderma harzianum* strain VOT1; QLT: *T. harzianum* strain QLT; BKNT: *T. harzianum* strain BKNT; VOB1: *Bacillus subtilis* strain VOB1; VOB2; *B. subtilis* cepa VOB2. Data inside the columns with different letters indicate significant differences in the Tukey test ($p \leq 0.05$). Dat = days after transplant.

Our results indicate that the application of strains of microorganisms as biological control products (in particular, *T. harzianum*) for the control of the downy mildew of cucurbits can increase the amount of 'premium' quality fruits by approximately 15 additional t per hectare. The results obtained in this study seem to coincide with previous studies' findings, in which some strains of *Trichoderma* improved the performance of several horticultural crops [19]. A similar study using cucumber plants treated with *T. harzianum* also produced cucumbers with higher contents of soluble carbohydrates, soluble protein, and vitamin C compared to the untreated plants, which correlates directly with a higher quality fruit [25].

The increased yield of cucumber plants and improvements in fruit quality could be related to the beneficial microorganism-plant relationship that occurs when *Trichoderma* invades the plant rhizosphere. This beneficial interaction is associated with the enhancement of plant growth by the microorganism and an increase in systemic resistance [26–28]. The fungus produces auxins to facilitate fungal colonization and increases plant nutrient uptake. These changes in the metabolism of the crops enhance productivity and fruit quality [29–31].

Applications of *Trichoderma* increased fruit yield of cucumber plants in treated crops even though disease control may not be as efficient. In our study, an increase in fruit production was observable in the plants treated with the *T. harzianum* VOT1 even though this treatment was not the best for disease control. These effects could be related to the secretion of harzianic acid (HA) and 6-pentyl-a-pyrone (6PP) as significant secondary metabolites by *T. harzianum*. These compounds directly enhanced fruit production in different crops, resulting in higher quality fruit with an increase in fruit size [32]. Yield improvements could also be related to an increment in the synthesis of volatile organic compounds (VOC), which are lipophilic compounds of low molecular weight and may act as promotors of plant growth [33–35].

T. harzianum strains may also improve the uptake of plant nutrients, with an enhancing effect on the efficiency of nitrogen use of the crop. This effect improves photosynthetic efficiency, which might also contribute to the increment in fruit yield and quality in cucumber plants treated with VOT1 when compared to the crops treated with *B. subtilis* [36–38].

4. Conclusions

Foliar applications of native strains of *T. harzianum* (VOT1) and *B. subtilis* (VOB1 and VOB2) can be considered viable alternatives for the control of downy mildew of cucurbits, as they provided better control than other commercial products, including Serenade Max®, Apolo®, and hydrogen peroxide. The best strains of microorganisms for the control of downy mildew of the cucurbits are the *Bacillus subtilis* strains VOB1 and VOB2. The VOT1 strain of *T. harzianum* provides adequate control over the disease and induces the highest yield of cucumber plants in comparison to the other strains. Further research is recommended to identify the mechanism by which *T. harzianum* enhances fruit yield and quality.

Author Contributions: Conceptualization, H.G.N.-P., B.E.O.-A. and I.E.-V.; methodology, H.G.N.-P., B.E.O.-A. and V.O.-P.; validation, J.I.V.-B., H.G.N.-P. and B.E.O.-A.; formal analysis, J.I.V.-B.; investigation, G.M.L.R.-A. and V.O.-P.; data curation, L.F.R.-S.; writing—original draft preparation, I.E.-V. and H.G.N.-P.; writing—review and editing, M.H.-C. and J.I.V.-B.; visualization, N.C.-H.; supervision, B.E.O.-A.; project administration, H.G.N.-P.; funding acquisition, H.G.N.-P. All authors have read and agreed to the published version of the manuscript.

Funding: This research received no external funding.

Data Availability Statement: Not applicable.

Acknowledgments: All authors and individuals in this section have consented to the following acknowledgments. The authors would like to thank "Invernaderos Arca S.P.R. de R.L. de C.V." for the greenhouses provided for the development of the project, specifically to Lic. Carlos Arturo Ramírez, owner and CEO; to Química Lucava S.A. de C.V. who provided the QLT strain of *Trichoderma harzianum* (http://grupolucava.com, accessed on 4 July 2015); and to Biokrone S.A. de C.V. for the donation of the BKNT strain of *Trichoderma harzianum* (http://www.biokrone.com, accessed on 21 August 2015).

Conflicts of Interest: The authors declare no conflict of interest. The funders had no role in the design of the study; in the collection, analyses, or interpretation of data; in the writing of the manuscript; or in the decision to publish the results.

References

1. Jaime, M.; Lucero, J.M.; Sánchez, M. *Innovación Tecnológica de Sistemas de Producción y Comercialización de Especies Aromáticas y Cultivos Élite en Agricultura Orgánica Protegida con Energías Alternativas de bajo Costo: Inteligencia de Mercado de Pepino*; Centro de Investigaciones Biológicas del Noreste S. C.: La Paz, Mexico, 2012.
2. SIAP. Sistema de Información Agropecuaria y Pesquera. Servicio de información Agropecuaria y Pesquera. 2016. Available online: http://www.siap.gob.mx/agricultura-produccion-anual/ (accessed on 1 May 2020).
3. Thomas, A.; Carbone, I.; Choe, K.; Quesada-Ocampo, L.M.; Ojiambo, P.S. Resurgence of cucurbit downy mildew in the United States: Insights from comparative genomic analysis of Pseudoperonospora cubensis. *Ecol. Evol.* **2017**, *7*, 6231–6246. [CrossRef]
4. Thomas, A.; Carbone, I.; Cohen, Y.; Ojiambo, P.S. Occurrence and distribution of mating types of Pseudoperonospora cubensis in the United States. *Phytopathology* **2017**, *107*, 313–321. [CrossRef]
5. Holmes, G.; Ojiambo, P.S.; Hausbeck, M.; Quesada-Ocampo, L.M.; Keinath, A.P. Resurgence of cucurbit downy mildew in the United States: A watershed event for research and extension. *Plant Dis.* **2015**, *99*, 428–441. [CrossRef]
6. Runge, F.; Choi, Y.; Thines, M. Phylogenetic investigations in the genus Pseudoperonospora reveal overlooked species and cryptic diversity in the *P. cubensis* species cluster. *Eur. J. Plant Pathol.* **2011**, *129*, 135–146. [CrossRef]
7. Cohen, Y.; Whener, T.C.; Ojiambo, P.; Hausbeck, M.; Quesada-Ocampo, L.M.; Lebeda, A.; Sierotzki, H.; Gisi, U. Resurgence of Pseudoperonospora cubensis the agent of cucurbit downy mildew. *Phytopathology* **2015**, *105*, 998–1012. [CrossRef]
8. Savory, E.A.; Granke, L.L.; Quesada-Ocampo, L.M.; Varbanova, M.; Hausbeck, M.K.; Day, B. The cucurbit downy mildew pathogen Pseudoperonospora cubensis. *Mol. Plant Pathol.* **2011**, *12*, 217–226. [CrossRef]
9. Lebeda, A.; Cohen, Y. Cucurbit downy mildew (Pseudoperonospora cubensis)—Biology, ecology, epidemiology, host-pathogen interaction and control. *Eur. J. Plant Pathol.* **2011**, *129*, 157–192. [CrossRef]
10. Neufeld, K.N.; Keinath, A.P.; Ojiambo, P.S. A model to predict the risk of infection of cucumber by Pseudoperonospora cubensis. *Microb. Risk Anal.* **2017**, *6*, 21–30. [CrossRef]

11. Thomas, A.; Carbone, I.; Lebeda, A.; Ojiambo, P.S. Virulence structure within populations of Pseudoperonospora cubensis in the United States. *Phytopathology* **2017**, *107*, 777–785. [CrossRef]
12. Call, A.D.; Criswell, A.D.; Wehner, T.C.; Klosinska, U.; Kozik, E.U. Screening cucumber for resistance to downy mildew caused by Pseudoperonospora cubensis (Berk. and Curt.) Rostov. *Crop Sci.* **2012**, *52*, 577–592. [CrossRef]
13. Hewitt, H.G. *Fungicides in Crop Protection*, 1st ed.; CABI Publishing, CAB International: Oxon, UK, 1998.
14. Olalde, V. *Centro de Investigaciones Avanzadas*; Personal Communication: Irapuato, Mexico, 2013.
15. Shoresh, M.; Harman, G.E. The molecular basis of shoot responses of maize seedlings to Trichoderma harzianum T22 inoculation of the root: A proteomic approach. *Plant Physiol.* **2008**, *147*, 2147–2163. [CrossRef]
16. Heydari, A.; Pessarakli, M. A review on biological control of fungal plant pathogens using microbial antagonists. *J. Biol. Sci.* **2010**, *10*, 273–290. [CrossRef]
17. Perazzolli, M.; Moretto, M.; Fontana, P.; Ferrarini, A.; Velasco, R.; Moser, C.; Delledonne, M.; Pertot, I. Downy mildew resistance induced by Trichoderma harzianum T39 in susceptible grapevines partially mimics transcriptional changes of resistant genotypes. *BMC Genomics.* **2012**, *13*, 660. [CrossRef]
18. Steiner, A. The universal nutrient solution. In Proceedings of the Sixth International Congress on Soilless Culture. International Society for Soilless Culture, Lunteren, The Netherlands, 29 April–5 May 1984; pp. 633–649.
19. Martínez, B.; González, E.; Infante, D. Nuevas evidencias de la acción antagonista de Trichoderma asperellum Samuels. *Rev. Prot. Veg.* **2011**, *26*, 131–132.
20. Ruiz-Sánchez, E.; Tún-Suárez, J.M.; Pinzón-López, L.L.; Valerio-Hernández, G.; Zavala-León, M.J. Evaluación de fungicidas sistémicos para el control del mildiú velloso (*Pseudoperonospora cubensis* Berk. & Curt.) Rost. en el cultivo de melón (*Cucumis melo* L.). *Rev Chapingo Ser Hortic RCSH* **2008**, *14*, 79–84.
21. Horsfall, J.G.; Barrat, R.W. An improved grading system for measuring plant disease. *Phytopathology* **1945**, *35*, 655.
22. Zeriouh, H.; Romero, D.; García-Gutiérrez, L.; Cazorla, F.M.; Vicente, A.; Pérez-García, A. The iturin-like lipopeptides are essential components in the biological control arsenal of Bacillus subtilis against bacterial diseases of cucurbits. *Mol. Plant Microbe Interact.* **2011**, *24*, 1540–1552. [CrossRef]
23. Elad, Y. Biological control of foliar pathogens by means of Trichoderma harzianum and potential modes of action. *Crop Prot.* **2000**, *39*, 709–714. [CrossRef]
24. Soria, F.M.J.; Suárez, J.M.T.; Trejo, R.J.A. *Tecnología para Producción de Hortalizas a Cielo Abierto en la Península de Yucatán*; SEP-DGETA, Centro de Investigación y Graduados Agropecuarios, Instituto Tecnológico Agropecuario no.2: Conkal, México, 2000.
25. Mei, L.I.; Guang-shu, M.A.; Hua, L.; Xiao-lin, S.U.; Ying, T.; Wen-kun, H.; Jie, M.E.I.; Xi, J. The effects of Trichoderma on preventing cucumber fusarium wilt and regulating cucumber physiology. *J. Integr. Agric.* **2019**, *18*, 607–617. [CrossRef]
26. Harman, G.E.; Howell, C.R.; Viterbo, A.; Chet, I.; Lorito, M. Trichoderma species-opportunistic, avirulent plant symbionts. *Nat. Rev. Microbiol.* **2004**, *2*, 43–56. [CrossRef]
27. Keswani, C.; Mishra, S.; Sarma, B.; Singh, S.; Singh, H. Unraveling the efficient applications of secondary metabolites of various Trichoderma spp. *Appl. Microbiol. Biotechnol.* **2014**, *98*, 533–544. [CrossRef]
28. Shoresh, M.; Harman, G.E.; Mastouri, F. Induced systemic resistance and plant responses to fungal biocontrol agents. *Annu. Rev. Phytopathol.* **2010**, *48*, 21–43. [CrossRef]
29. Contreras-Cornejo, H.A.; Macías-Rodríguez, L.; Cortés-Penagos, C.; López-Bucio, J. Trichoderma virens, a plant beneficial fungus, enhances biomass production and promotes lateral root growth through an auxin-dependent mechanism in Arabidopsis. *Plant Physiol.* **2009**, *149*, 1579–1592. [CrossRef]
30. Grossmann, K. Auxin herbicides: Current status of mechanism and mode of action. *Pest Manag. Sci.* **2010**, *66*, 113–120. [CrossRef]
31. Lombardi, N.; Caira, S.; Troise, A.; Scaloni, A.; Vitaglione, P.; Vinale, F.; Marra, R.; Salzano, A.; Lorito, M.; Woo, S.L. Trichoderma applications on strawberry plants modulate the physiological processes positively affecting fruit production and quality. *Front. Microbiol.* **2020**, *11*, 1364. [CrossRef]
32. Pascale, A.; Vinale, F.; Manganiello, G.; Nigro, M.; Lanzuise, S.; Ruocco, M.; Marra, R.; Lombardi, N.; Woo, S.L.; Lorito, M. Trichoderma and its secondary metabolites improve yield and quality of grapes. *J. Crop Prot.* **2017**, *92*, S11–S12. [CrossRef]
33. Garnica-Vergara, A.; Barrera-Ortiz, S.; Muñoz-Parra, E.; Raya-González, J.; Méndez-Bravo, A.; Macías-Rodríguez, L.; Ruiz-Herrera, L.F.; López-Bucio, J. The volatile 6-pentyl-2H-pyran-2-one from Trichoderma atroviride regulates Arabidopsis thaliana root morphogenesis via auxin signaling and Ethylene Insensitive 2 functioning. *New Phytol.* **2015**, *209*, 1496–1512. [CrossRef]
34. Lee, S.; Yap, M.; Behringer, G.; Hung, R.; Bennett, J.W. Volatile organic compounds emitted by Trichoderma species mediate plant growth. *Fungal Biol. Biotechnol.* **2016**, *3*, 7. [CrossRef]
35. Paul, D.; Park, K.S. Identification of volatiles produced by Cladosporium cladosporioides CL1, a fungal biocontrol agent that promotes plant growth. *Sensors* **2013**, *13*, 13969–13977. [CrossRef]
36. Benítez, T.; Rincón, A.M.; Limón, M.C.; Codón, A.C. Biocontrol mechanisms of Trichoderma strains. *Int. Microbiol.* **2004**, *7*, 249–260.
37. Kumar, S. Trichoderma: A biological weapon for managing plant diseases and promoting sustainability. *Int. J. Agric. Vet. Sci.* **2013**, *1*, 106–121.
38. Sood, M.; Kapoor, D.; Kumar, V.; Sheteiwy, M.S.; Ramakrishnan, M.; Landi, M.; Araniti, F.; Sharma, A. Trichoderma: The "secrets" of a multitalented biocontrol agent. *Plants* **2020**, *9*, 762. [CrossRef]

MDPI AG
Grosspeteranlage 5
4052 Basel
Switzerland
Tel.: +41 61 683 77 34

Horticulturae Editorial Office
E-mail: horticulturae@mdpi.com
www.mdpi.com/journal/horticulturae

Disclaimer/Publisher's Note: The title and front matter of this reprint are at the discretion of the Guest Editors. The publisher is not responsible for their content or any associated concerns. The statements, opinions and data contained in all individual articles are solely those of the individual Editors and contributors and not of MDPI. MDPI disclaims responsibility for any injury to people or property resulting from any ideas, methods, instructions or products referred to in the content.